TO THE PEACE AND BACK

ISBN: 978-0-9846536-1-4
Library of Congress Control Number: 2012931963
Published by Global Authors Publications

Filling the GAP in publishing

Edited by Barbara Sachs Sloan
Cover Art by Kurt C. Rolle
Interior Design by KathleenWalls
Cover Design by Kathleen Walls

Printed in USA for Global Authors Publications

TO THE PEACE AND BACK

BY

KURT C. ROLLE

DEDICATION:

To Joy, Who made it all happen

AKNOWLEDGEMENT:

I wrote this book mostly to tell some of the happenings of a short two year span of time involving my family and friends. I want to thank all of my family, my six children, and many dear friends who helped make this event happen. In particular, Gary Berning and his family in Bradford, Ohio, Jesse Peachey and his family in Baytree, Alberta, George Blakus in Baytree, Alberta, Ian Marchuk in Baytree, Alberta, and many other people too numerous to mention.

I want to thank Kathleen (Katy) Walls who seemed to have early on recognized in the manuscript an interest for readers. Her encouragements and guidance made the book possible for publication. Also, Barbara Sachs Sloan did an expert editing of the rough manuscript, not only mechanically correcting it but making crucial suggestions that certainly improved the book. Thank you both.

And always I want to thank everyone who I have met while trying to grapple with reality. It is, finally, all of us together that makes life rich, including those who are antagonistic or seem to be totally against us.

Finally, I want to thank my dear animals who were involved in the adventure; Bess, Morning Star, Chief, Flossie, Prissie, Friendly, and all the others.

CHAPTER ONE
The Idea and the Dreams

August in Ohio is a time of upcoming harvests, warm days and nights, and the hints of autumn as well. In eastern Darke County, where our small farm was located, the August of 1978 was a busy time for us. We had set in motion a plan to move to the Peace River Country of northern British Columbia and Alberta. And this all started long before this time. As I think about it all and reflect again, it started a long, long time ago for me....

I grew up from the age of three in east central Indiana during the Second World War years and through the Korean War and on. And I am not sure what makes us do things, but it is surely partly our environmental experiences and partly our genes, bloodlines, evolution if you will, and maybe nature or the Gods or whoever is pulling our strings. But I was born on the shores of Lake Michigan in northeastern Wisconsin, in a small town called Algoma, meaning "Park of Flowers" in the Potawatomi language. Then again, the town was named Ah-Ne-Pe by the Potawatomi Indians, meaning great gray wolf. I like that better. When I was three years old, in the summer of 1941 it was, my family moved to Indiana because my father had an offer for a good position as a pattern and model maker at the Overmyer Mould Company in Winchester, Indiana. I say it was good because it must have paid him enough for him and, besides, he had a precarious position at a church furniture factory in Kewaunee, Wisconsin, 15 miles south of Algoma. If you have a feel for history, remember that the summer of 1941 was just a couple of months short of December 1941 when the war became the Second World War, and I am not sure that my father could have moved to Indiana after that awful event. All of my family were landed immigrants from Germany, and Germany was now the enemy. However, he was an accomplished wood carver with not only the mechanical skills to carve wood but seemed to have ideas on creative things, an artist so to speak.

Overmyer furnished iron moulds for the glass industry; the moulds were used to blow glass bottles and jars of all sizes and shapes. The regular moulds for common bottles were fairly straightforward to make. The patterns (thus the term pattern maker) were made of wood, and then iron castings were poured in a sand cavity formed with the wooden patterns. A fancy or irregular pattern required that a model be made first, and I remember that whenever a difficult or fancy bottle mould or pattern needed to be made (usually whiskey or brandy bottles) it was my father who did the wooden model and the pattern as well. His wood carving trade and skills served him well.

My grandfather, my *opa*, was also an accomplished wood carver. He was from the old school. He tried to create works that had lots of detail, lots of curved and sensuous figures, like nature would have it. He was also an accomplished wood inlay craftsman. So too was my father. And *Opa* became an anachronism; our society continues to demonstrate the principle of increasing entropy. We want everything to be cheap, ugly and easy for us. Oh, no one says that, but look around at the hideously designed and constructed buildings we build now; the music (if that is really what it is) and the literature are mostly irrelevant and gross. Wood carving became an anachronism because first of all there were copy machines that could "duplicate" a Venus de Milo a thousand times; just keep the tools sharp. But finally all of our surroundings didn't need edifying oak leave clusters, beautiful nudes holding a vase, or music to make one actually feel reflective. There will always be some who actually do some substantive studying and work, but the vast majority of our citizens is following like a plow behind a tractor, or maybe like the gas chamber operators just follow orders.

Anyway, my opa left his native Deutschland in around 1928, pulling his wife, my oma, and his three children--my father, Tante Mimi (Wilhelmena), and Tante Marie--and came to America for a new life. But even then he was an anachronism because he was looking for freedom and a country not enamored with war and the military. He told me a number of times that he disliked the German Kaiser and his policies, but he had also experienced war on the eastern front, in Poland and western Russia, and he absolutely hated war of any kind. So he wanted to come to America which was a peace-loving and free country.

His new country, as it turned out, had neither of these admirable traits. Again, no one would admit to it, but his new country spends more money by about a factor of two or more than any other country on war and weapons of mass destruction. And freedom is surely illusory when one looks at the conniving for political power with the ridiculous money spent for "electing" someone. Substance has become nothing but only money. The country claims to be one which "Trusts in God," but the gods they trust are those despots pictured on the various paper money bills. And "under God" is certainly a good description of the fact that the country is under the control of money, the bigger the better.

So my opa arrived in 1928 in New York City, found a job as a woodcarver in Chicago and settled in. One year later my father and the rest of the family moved over, and they lived in Chicago to follow their dreams. Their immediate family, including brothers, sisters, cousins, aunts and uncles, remained in Deutschland, lamenting the fact that Opa had forsaken the fatherland. In retrospect I think he didn't forsake his country, but his country forsook him. I remember even during the war years when to be a German was a sin, at least in the USA, he received and read the newspaper *Die Welt am Sonntag*, *The World on Sunday*. It was all German and sent to him by a dear relative of mine who was in Deutschland and whom I don't know. I sometimes looked through it although I couldn't at the time read, let alone speak, German. But it was not Deutschland that my opa forsook; he loved it, as I know he loved the USA. The war, with all the political corruption, and not just on the German side, and the striking up of a second world war must have made his heart soo heavy.

Anyway, my father was more pragmatic; his woodcarvings were always more straight formed, modern or so. Not ugly but not sensuous or natural either. I remember being around when my father and my opa would sometimes discuss art, woodcarving, and the crafts. The discussions were not heated, but it was clear that Opa was not at all convinced that straight lines were anything other than a short interlude for a beautiful flowing ogee curve or cusps between smooth, sensual mounds. My father was from the modern school and knew better than Opa what was happening, or at least he was willing to follow. Interesting, because it was my opa who moved to America, not my father, who only followed.

Once in the US, however, things evolved. Actually, I believe that my Opa too knew what was happening, but I suspect that he wanted to change it. But he was, as I said, an anachronism. So my opa retired in the late 1930s, under the new deal Social Security legislation and managed to eek out a living with his monthly check. It must not have been much, but at any rate Tante Marie and Tante Mimi had plans. Actually my Uncle Herman, husband to Tante Mimi, had the plans. He would farm in Sawyer County, Wisconsin, where the land was cheap and fertile. He would milk cows in the Dairy State, and it would be good. Uncle Herman was a sailor for the Kaiser and came with Tante Mimi to the US, to follow his love. They never had children but were decent people who could have raised children better than many people I know. During his stay in Chicago, while my father and Opa were woodcarving, he did mechanic work and some entrepreneurial stuff with oil-burning furnaces. He seemed to have a kind of talent for tinkering, but nothing really came of it all. He did, though, move to northern Wisconsin, in Sawyer County near the town of Lemington around 1940. It was, when I remember it in the 1940s, beautiful country. Except if you looked closely you could see the rape that occurred not fifty years before when the robber barons and lumber barons came and clear cut the virgin white pine forests. So Uncle Herman and Tante Mimi purchased about 200 acres of rough Sawyer County land on County Trunk C, a main road from Lemington north to State Highway 27 and 70. County C was always gravel as I remember it. The land was about two miles north of Lemington and had a gravel pit right at the road and directly across from the homestead they built.

At the same time, Tante Marie was working in Chicago as a bookkeeper, an accountant I think, and had two children, Clark and Elaine. Clark was seven years older than me, and Elaine was seven months older. And Tante Marie purchased a 40-acre parcel adjacent to Uncle Herman and Tante Mimi's place, probably with the intention of moving up eventually but for the interim having Opa and Oma build her a home, live there free rent, and take care of Clark and Elaine.

My first experiences of this venture were around 1944 when my family went to Lemington in the summer on the Soo Line train from Chicago. I vaguely remember boarding a Pennsylvania passenger train in Winchester at the depot and riding to Fort Wayne where we must have caught another train into Chicago. I do remember seeing the steam engine coming into the station in Winchester under a load huffing and puffing with a beautiful smoke cloud coming from the smoke stack. And when the engine passed by, the ground shook. I think I raced into the railroad station building as it was all too overwhelming. But we must have stayed in Chicago with my Tante Marie for a day or so before catching the Laker, the Soo Line daily that left Chicago in the late afternoon, and arriving in Duluth the following morning. We would get off at Lemington, a flag stop, in the early hours of the morning. I remember boarding the train at a smaller suburban station outside Chicago, and the train was packed. We had to sit in the vestibule, the interval at the end of each car where one actually came into the train. The windows were open because it was very warm; the trains at that time had no air conditioning. Somewhere along the first 100 miles a puff of soot from the steam engine powering the train came into the vestibule and a bit of ash got into my eye. That was uncomfortable and bothered me for a few hours, but being young it went away. My love for train riding did not go away... And I can still see that first arrival, or maybe it was the second one the next year. It was early morning, the sun had come up and it was austere and damp. The road, County C, crossed the tracks here, and my Tante Mimi was waiting for us with her Willys sedan. My oma was probably along with Tante Mimi as our train chugged to a stop, and it was cool for middle June. County C made a sharp left turn at the tracks, heading north and up a small hill. To the right there, overlooking the tracks and Lemington, was a stone house and a neat set of buildings. A round stone silo, probably the only one in Sawyer County, stood next to a wooden bank barn. This turned out to be Jack and Mary Eckel's place, good friends to Tante Mimi and Uncle Herman. The train jerked some as it slowly accelerated but increased in speed to the north, and finally while leaving, the village of Lemington could be seen. It was a hamlet; as one looked southwest down the road there were three houses on the right adjacent to one another, a patch of scrub woods and after another 150 meters or so a white general store. The general store was general as it had a single hand gas pump in front, a motley selection of groceries and the village post office. There were some hardware items to be had there and also a few clothes, but not much. It was a two-story building and more long than wide. The roof was gabled, and I can't remember but there was probably a false front, the classic square wall at the front to give an impression of grandeur. On past the store maybe another 200 meters was the Lemington School, a nice one-room square white wooden building where my two cousins went. A stub road went north just past the school to a farm up in the hills to the north. It was a dead-end road, really only a lane for that farm, which was not to be seen from County C, the main and only real street in Lemington.

The south side of the road had no buildings at that time until just across from the school where there were three more houses. On past those houses maybe 300 meters was a sawmill that was served by a stub siding of another railroad, the Omaha, which crossed the road at about a 45-degree angle, heading west and north but then circling to the southwest to Rice Lake, Wisconsin. I can't remember what was used to drive the sawmill, but it was not pretentious, an open shed roofed structure over the main saw and carriage. The Omaha railroad circled past Lemington on the south and crossed over the Soo Line just south of the town. We had traveled under the overpass when we approached Lemington.

As we headed to the farmstead in Tante Mimi's Willys, we rattled and bumped over County C. There had been some good rains, and the road was just about gone, with washouts. There was a severe hill to climb not 1 kilometer from Lemington. It was a south-facing hill, and the road went straight up, but with the meager gravel and ruts, it was a struggle to reach the top. We did, though, and another two kilometers on a straight rutted gravel path we reached the home or homes. The setting was all on the west side of the road, to the left. First was Uncle Herman and Tante Mimi's one-room home. It was a raw clapboard-sided structure, inside kind of divided with a curtain to make two rooms. There was a thin-walled box stove that served to heat the house, and as I remember, the nights were almost always cool enough for a fire.

There was a small kerosene stove they must have used for cooking.

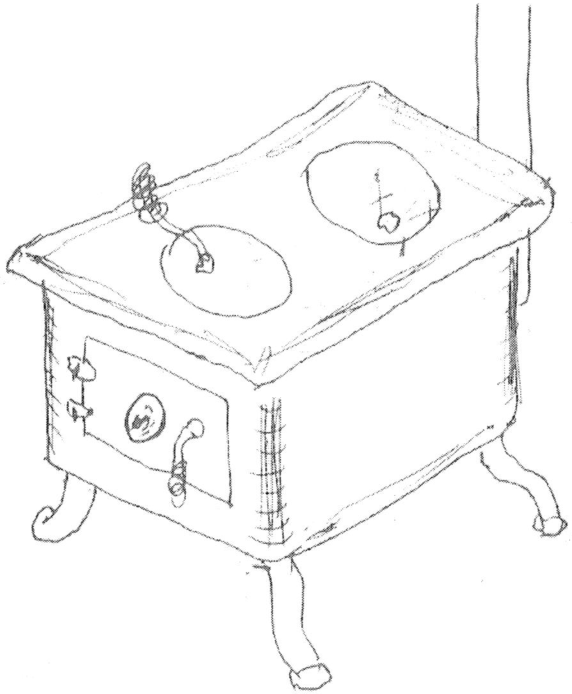

To the north of the house was a bank barn, gambrel-roofed, with roll roofing, layered red and green. Looking back I thought the barn was huge, but it was probably 25 feet wide at most and maybe 50 feet long. The short gabled side faced the road, and at the back was a door to the basement so Uncle Herman's cows could get in. The foundation was stone up to the main floor superstructure. From there up it was a framed building. In the following years Uncle Herman would add on to the front to extend it out for a milk house and more hay storage. I think he always used loose hay and sometime he installed a hayhook on a track at the peak of the roof inside the barn to pick hay from his loaded hay wagon, which by the way was really a single-axled trailer but built substantial enough to have a good-sized hay rack, and bring it into the barn. The hay hook was like a big dual set of tongs, looking like a set of teeth and which literally bit a clump of loose hay from the wagon and then raised it to the ridge before moving it into the barn to be deposited in the mow. My cousin Elaine and I often were assigned the task of tromping the hay down after a bite had been discharged from the wagon into the loft. Loft work connotes a hot summer day with sweaty, cramped conditions in a dark hot mow, but I remember it to not be hot or cramped. Sawyer County was not hot anywhere at that time, and we surely enjoyed doing the tramping down. I am not sure that Uncle Herman, Tante Mimi and my cousin, Clark, who was a main help for Uncle Herman, didn't just put us up in the mow to get us out of the way.

Then, to the north of the barn was Opa and Oma's house. At the first visit to Wisconsin this was a framed building and about halfway up around the building was stone, mortared together. The stones were nice smooth granite-like multicolored rocks of varying sizes. But it presented a nice interesting, solid surface with lots of curves and a few minor crevices. The house had five rooms plus a small dressing room on the main floor and a full basement with Opa's workbench and, over in a far corner, a stool. They never had electricity or a telephone while I was there for summers. Over the following years, Opa would complete the outside of the house with nice smooth stones up to the roof. The stones were all picked from the fields right there. And I remember picking stones from the fields in Sawyer County was a continuing chore for Uncle Herman and sometimes Opa. The region was, supposedly, overridden with glaciers, and lots of the top soil was scraped south. So the land there had lots of granite stones.

The running water for the two houses and for the barn came from a hand pump near Opa and Oma's house. It was a polished red iron pump with a long handle. A pipe traversed a gully separating the pump from the barn, and this supplied water to the cow drinking trough at the barn along with another tank to keep milk cool, as if that was a problem. Water for the houses was carried by white enameled 10 Liter (three gallon) bucket from the pump to the houses. Tante Mimi's water was a bit of a chug to carry, but Oma and Opa's water was just 20 meters or so up a short flight of stairs into the house. Oma and Opa had two buckets of water that were placed on a low shelf in the kitchen behind a door. A metal long-handled ladle was always at the bucket for drinking fresh water. It took me a short time to get used to drinking out of a common ladle, but once I was used to it, all seemed so natural and hygienic too.

And the lights were all kerosene lamps or Aladdin lamps. The Aladdin lamp was a pressurized gas lamp that gave off a much brighter light, but the kerosene wick lamps were more comfortable. I remember sitting at the dining room table in Oma and Opa's playing Pinochle or working a jigsaw puzzle with a kerosene lamp keeping enough light for us.

We would return every summer to Wisconsin until I was maybe 12 years old, 1950, and I hope to eventually write more of those times. For the first few times we always rode on the Soo Line. Later on, when my father was able to purchase a more reliable automobile, we drove up. Driving in a car was no fun compared to the adventures and comfort of a train, but my father was obsessed with cars, even to the point of terrorizing us with the chances he took on crowded highways. I would have always preferred doing the trains.

So my early experiences made fond memories and left lasting imprints on me. I thought then and do now that the sociology and technology of it all were at a higher plane than the more modern dilemmas. And I too must be an anachronism, like Opa, because of this, whether by environment or evolution or heredity. But I know too that pragmatism, economics, and, yes, even religion, are but passing attitudes and events we need to endure. Perhaps the only real people, *Human Beings* (*Humane Beings*), have been and are the native Indians of old and who have embraced their real heredity, the aborigines who have resisted encroaching civilization, and the African and American Negro who must have a real difficulty finding their true heredity but in many cases are enduring. Even the Europeans, Asians and others can become *Human Beings*, but the mantles of civilization have obliterated much that must have been real and good. Can heredity span a 5,000-year chasm of progress and human slavery to give a person the option of being a *Human Being*? My answer to this question has always been, "We need to try, or *Tätig zu sein ist des Menschen erste Bestimmung.*"

So as I grew up, I needed to do something. And my interest was always with railroads: designing, building and operating steam engines. What else was a railroad if not the steam engine? And if one studies the technology of it, the genius that came from England for a steel wheel rolling on a steel rail (actually the wheel does not roll on a surface, but that is another story for another time) is probably the highest point of vehicle designs, outside of walking, or maybe riding with a horse. It is almost sustainable. So, upon graduating from Winchester High School in 1956, I went to Purdue University to study mechanical engineering. Purdue was a center for railroad steam engine testing, the boilermakers. And I had four years of academic experiences with mechanical engineering, graduating in 1960 and just barely. My class was the first at Purdue to require calculus, analytic geometry, and nuclear physics for graduation. But I persevered and, looking back, Purdue was at that time a good school.

It was 1960, Eisenhower was president and the economy was not good. (Has it ever been good, particularly from the standpoint of the political groups not in power at the time??). Engineering positions were not plentiful, but somehow I was fortunate in having acquired a summer job with the Perfect Circle Corporation during 1959, making a good enough impression with my technical skills, and making some friends so that they offered me a permanent position designing and making piston rings for internal combustion engines. Well, this was not steam engines and not even railroads, but that too is another story for another time. So once again I kept at it, and I took the position. After maybe three years it became clear to me that there had to be something better than a 40-hour week of making piston rings, valve seals and cylinder sleeves. And not taking any vindictive swipes at anyone, the people at Perfect Circle were civil, doing their jobs and following orders. I recalled from time to time my experiences in northern Wisconsin and wondered if Canada was a better place to live and endure. It was still wild in much of the territory, not imprinted with civilization. And besides, it was not that far away and seemed like a viable option.

I did meet and marry my wife of now 50 years, Joy, in 1961, She was the most wonderful person I had met, she was receptive to anything I wanted or said, she told me how great I was, and we had a good relationship. We had a handsome son after 11 months of marriage, Kurt, and a beautiful daughter after another two years, Lorelei. They have been joys to my life ever since they came, and I remember often thinking where did they come from. It surely was not me or Joy who was capable of bringing two human beings into the world.

And then, in 1964, I decided to do graduate work in mechanical engineering. After all, everyone said education was the most important thing you could do. Education separated us from the "lower" animals and gave us license to take anything and everything from the earth. Even the Yahwe, God or Allah told us that the earth is there for us and that we should get lots of material goods and material wealth. Ambrose

Bierce would know and appreciate my thoughts on this take. So I enrolled in an advanced calculus course at the University of Dayton (UD), in Dayton, Ohio. My undergraduate record is not good, barely acceptable, but my motivations now were stronger, and I now had a wife and two children, so I applied myself and earned a solid A in the course. I remember that mathematics (advanced calculus) was supposedly a rational, deductive subject. But the course was a rote learning course. I needed to memorize definitions, lemmas, and theorems. So I did, and it was predictable; I regurgitated all that I memorized and earned an A, the best. I had gone through four years of undergraduate schooling and always just getting by because I didn't want to just memorize things; I wanted to learn, and often times that was irrelevant to the subject matter. So my grades were Ds, Cs and an occasional B.

Perfect Circle had a printed policy of paying for any tuition for graduate credits earned, with a C or better. When I told my supervisor, Marvin Mahoney, that I had earned an A in advanced calculus, he just said they didn't need advanced calculus in the Piston Ring business and they wouldn't pay for it. I never argued with him, but right then I learned positively something that had festered in me for a long time: Businesses have no ethics other than their gods, money and financial profits. So I knew then, positively, that I needed to get on with my life and get away from Perfect Circle. Besides I was going to take another course at UD, and Dayton was 80 kilometers from Richmond, Indiana, where we lived. Joy was all for me.

So I took a position (actually this was a job) with Monarch Machine Tool Company that had an engineering office in Dayton. The main plant was in Sidney, Ohio. They made machine tools--engine lathes, drill presses and milling machines--and they were a highly respected company. Well, I found out quickly that they may have been highly respected, but their respect was coming from corporate America, just what I wanted to bail out of. Their operations were based on maximizing profits and reducing costs for overhead (like engineering work?). I stayed for six months at Monarch Machine Tool Company, getting some good experience and learning in the process. I jumped to the federal government's Fuel Systems Group of Systems Engineering Division at Wright-Patterson Air Force Base. It was just east of Dayton, and since we had moved to Kettering while I was working at Monarch, everything was getting better. The Fuel Systems Group was headed by a gentleman named Clyde Schaffnit, and the group was responsible for monitoring fuel systems of the air force vehicles: cargo airplanes, bombers, fighter airplanes, and helicopters. Monitoring meant, I found out, writing military specifications that defined how such and such a system or piece of equipment needed to operate and in what sort of environment. This involved also defining testing programs, installation method, and service programs. In principle it also meant that we should actually design systems and their components. But I came to this group in 1966, and the government had become neutered or castrated, so to speak. Before and during the Second World War, the Air Force, at the Wright-Patterson base actually designed and built prototype aircraft and tested them. The base had a complete facility for machine, welding, casting, forging and fabricating aircraft. It probably was the most advanced manufacturing facility in the country. But with the enlightenment of post war years, the "private" free-enterprise, capitalistic people connived to make money for themselves. They wanted to do the designing, development, manufacturing, and testing of new Air Force equipment systems and vehicles. Boeing, Lockheed, Grumman, Sykorski, Bell, Northrup and many other companies became suppliers for Air Force systems and what profits they would get. This went on with all the government agencies such as the Army, the Navy, transportation, and on and on. Well, these companies wanted profits, not good reliable equipment, but that is another story too. So when I was at Wright-Patterson, there were large hangers or manufacturing areas that were abandoned because they were not now needed since private industry was doing the work. I often thought about why they called themselves private, free-enterprise companies when their only customer was the government and its the war departments, and they didn't even need to finally prove any quality control with their products. It was a perfect system for them, and today it turns out to be even more so. When I was at Wright-Patterson, in 1966-67, the theory, the military policy, required that everything be contracted competitively. This meant at least two vendors or manufacturers were needed to let a contract to build or supply something to the Air Force in this case.

Well, private companies managed to combine, go bankrupt, get taken over by a bigger company or whatever so that now there are mostly sole companies for any one item, aircraft, or service. Very nice for the businesses that have no competition and now no need to even do a good job. This is called a monopoly and suits businesses just fine. I found out one other interesting item about all this. During my stay at Wright-Patterson I became involved with analyzing and troubleshooting the fuel system of a large Sikorski transport helicopter, the HC-153. They were just then cranking up the Viet Nam conflict and were using these helicopters to ferry men and supplies to and from the battlegrounds. And the helicopters were being lost, many due to running out of fuel. This was a waste of a machine, but it seemed to me the worst part was the waste of human lives. A helicopter without fuel over a battle area does not allow good chances for the survival of the people in the machines, and they didn't survive. So the Air Force wanted to know why the fuel was running out at inopportune times. I remember it was myself and another engineer, Dexter Kalt, who were assigned to look into the whole fuel system of the helicopters. Well, the fuel system had no glaring deficiencies like too small a fuel tank, unvented fuel tanks or pumps, or other design defects. What was lacking, however, was a fuel gauge. Every automobile has a fuel gauge, and every other vehicle that is expected to travel any distance has a fuel gauge. But this multimillion-dollar helicopter, designed to transport men and equipment over significant distances and over hostile territory, had no such a device. We told our supervisors, the Air Force and Sikorsky that they needed a fuel gauge in the HC-153. We even took two trips to Stratford, Connecticut, to meet with the folks at Sikorsky. After some discussion it all came to a head and the senior senator from Connecticut, whom I cannot now remember, interceded and told the Air Force and us that it would be too expensive to install a fuel gauge. This was just another experience in how capitalistic, free-enterprise, private business operates. Patriotism is very cheap, and all one needs to do is wrap a flag around one's self and you are suddenly a patriot. Forget ethics or care for others. I know that the same sorts of things transpired in Germany under the National Socialists, and under the communists in Russia, and they are going on right now in the USA. Oh yes, in the USA there was a War Department during the Civil War, the Mexican War, the First World War, the Second World War, and the Korean War. After the Korean War their name changed to Defense Department, and if you don't know why that change was made, then you may as well put this book down and do something else. You would not understand it.

During my stay at Wright-Patterson I also engaged in designing an aerial refueling pod for helicopters and to be universal for all Air Force aircraft. It was done with paper and pencils and could today be done with a solid modeling computer program so others could understand it. I don't know what became of my drawings, but they probably were tossed away so a private company could make their profits bigger.

In the summer of 1967 I learned of an opportunity to teach mechanical engineering technology at the University of Dayton (UD). I was never comfortable speaking to groups of people, but as I mulled it over in my head, I realized I did have the interest in teaching and considered it a worthwhile way to serve people, unlike making piston rings, machine tools and war tools. I do remember during my second year of engineering at Purdue that I really wondered if I wanted to do engineering. I thought about transferring into teaching, maybe just going to Ball State Teachers College in Muncie, Indiana, and whatever. My sister told me not to do it. I remember her saying that anyone can do public teaching but not everyone can do engineering. Well, I bought it, not that I today believe that everyone can really teach nor that not everyone can do engineering, but I kept at the engineering at Purdue and did finish it. I don't ever regret that. So I did accept a position as assistant professor of mechanical engineering technology at UD. It turned out to be a good experience for me and much more rewarding than I would have ever expected.

Teaching mechanical engineering or engineering technology was a challenge for me. I wanted to present material that students could use in a wise way. I was not driven by the profit motive, and I took a salary cut to leave Wright-Patterson to join UD. Not only that, but the War Department was always going to be there and UD was having some budget problems. But all of my experiences at UD were good, really. They never came around talking about making more profits, they never said their bottom line was profit, and they never told me that such and such subject matter was not necessary for the good of UD. It did bother me that UD had a large research institute, devoted to research. Well, research began at UD with a contract from Wright-Patterson Air Force Base to digitize data. Wright-Patterson was a Strategic Air Command (SAC) Air Force base, having the largest fleet of B-52 bombers. These were the largest aircraft in the military, carried nuclear warhead (bombs), and flew every day from Dayton to the Arctic Circle. A movie, Doctor Strangelove, is a parody on this operation. Aircraft are made mostly of aluminum because it is lightweight, easy to form and join, and reasonably strong. It has a problem in getting tired, fatiguing, after a number of loadings and unloadings. A tired wing of a B-52 can fall off if it is too tired, and that could be catastrophic, particularly with a nuclear bomb in its bay. So the B-52 was equipment, wired with strain gauges to monitor the number of loading cycles of a particular aircraft and to give maintenance engineers and crews the ability to decide when to replace a certain tired wing. So there was lots of strain gauge data, and it was on strip charts, called analog data. Like a squiggly line on paper. Well, engineers want numbers, not squiggly lines, and taking this line and converting it to a number is called digitizing or analog to digital conversion (A/D conversion). In 1967 and even up until 1990 or so, manual conversion was the way to do it. And all it took was a person to look at the line and assign a number every short segment. It was tedious work but important to the Air Force. College students could do it well, and UD had lots of college students who needed money to pay for their tuition and beer. The research institute was formed at UD to handle this work, and it grew enormously from about 1950 until today UD has probably the largest research institute among the nation's Catholic universities. I thought it strange that a Catholic university wanted to engage in War Department research, but Defense Department research is quite another matter. Or is it??

I had a good friend at UD who worked in the research institute, Greg Williams, and with whom I had many discussions. In the 1968-1975 time period the Viet Nam conflict was splitting the country along some very serious lines. I remember it being a time when the whole War Department was coming under question. The conscripted soldiers took the brunt of the blame for the war as is always the case, and the generals, admirals, and the President were mostly insulated from it all, and it all never came to a French Revolution with a guillotine for deserving persons. Even college students were killed (Kent State), citizens were beaten in various demonstrations, and the private industry folks were getting uncomfortable. Their profits for weapons were being compromised. Oh they still got all their profits, but I know that they were getting uncomfortable knowing that probably one half of the country was actually questioning the policies of conducting the Viet Nam exchange. The war mongers wanted to just totally annihilate Viet Nam, destroy the country, in order to save it. They tried to. But even a small country like Viet Nam is hard to annihilate, and even the USA, with the largest military expenditure in the world, could not afford to do it. It was almost like an Alice in Wonderland absurdity, but it was not fiction. The result was that the demagogue president who in 1964 was elected on a "peace" platform with the largest plurality ever (I think) decided not to run for president in 1968. So we elected another demagogue who turned out to be just as bad and actually tried more seriously to obliterate Viet Nam. He too became bankrupt, although I think he was already morally and ethically bankrupt in 1968. By 1972 he demonstrated his bankruptcy and got elected again. That became a disaster for him and the country. He never finished his second term and in 1976 we had the hint of a real person occupying the White House. I was always able to give President Carter the benefit of the doubt, and I think he did what he thought was right and demonstrated some intelligence and moral principles. I still found everything very difficult to take and somehow felt that the President did not change a country's citizens. This lesson was doled out in early times, the Israelites, the Egyptians, the Hittites, the Greeks, the Romans, and even the French. But it never really gets learned.

During my period at UD I wrote a textbook on thermodynamics that has, over the years, been rather popular with some people. I had intended it for engineering and engineering technology students who may have struggled with higher mathematics but who could understand the ideas of the subject of thermodynamics. I have felt good about that effort and still do.

And I had kept somehow gravitating toward Canada. Canada was not directly in all the world's meddling although it always kowtowed to the USA. I knew that if I could just get away and kind of ignore it all, I would be better off. Joy felt the same way.

While we lived in Kettering, Ohio, we were blessed with two more lovely children, Timothy Jon and Heidi Anna, both born in the Kettering Memorial Hospital. This was owned and operated by Seventh Day Adventists. This is a religious group who believe in the resurrection of Christ, the imminent end of our present world or society, and the last Judgment. It seems to have been founded by a William Miller in the early 1800s. He announced that Christ's second coming would occur in 1843 and the end of the world would then occur. Mr. Miller lived until 1848 and he surely must have been a leader. His prediction was way off, I think, but somehow the sect kept enough members that it survived. I did not study their present doctrines and beliefs, but what impressed me about the hospital was that it was clean, seemed to be efficiently operated, and served vegetarian food. I partook of their vegetarian selections and avoided meat entirely while I waited in the hospital for Joy to give birth to both children. I remember getting cooked parsnips, something I hadn't had before, and I thought they were delicious. But the birth of these two children was the incomparable experience of having now four children in our family. It was just wonderful, and I know now that the most pressing problem of the world is overpopulation. I can't and wouldn't apologize for two more beautiful human beings in the world, and I have no answer yet to what, as a species, we should do. Both Tim and Heidi have given me more wealth, much more, than I deserve.

We had a few experiences while we lived in Kettering that prodded us to the country and simpler living. We had a plat, slab range-style home. It had brick veneer and a hipped or hip roof and had been built probably 30 years before we lived in it. It was a platted community within Kettering, the houses all built in short order. And while it looked nice, there were some real deficiencies with it. The window frames were steel, and in the winter ice formed inside the windows whenever the temperatures dropped below freezing outside. The steel frames were great for conducting heat, and in the summer the windows were hot. Drapes covered the windows so we didn't look out much, but that didn't really solve the problem of the steel frames. I am a mechanical engineer and should have known this was bad, but I learned it firsthand. It took the window manufacturers a bit longer to learn so that now wooden windows are back and otherwise thermal breaks are included in the frames. We had a forced air furnace that distributed hot air from the centrally located furnace to the house's outside walls, usually below the leaky windows. By the time the air circulated around the house the last vents gave only lukewarm air. We had a fireplace, a real one, which was not too big. I tried to keep the house warm in the cold weather by burning wood in the fireplace, and it helped. And our three oldest would snuggle up to the fireplace on cold evenings to get some of the heat. I had to get the wood by scrounging in one of the local parks, either Hills and Dales Park or Indian Ripple, where dead limbs often were lying around. It was not a good way to burn wood or get it. And looking back, this was middle Ohio, latitude of 36 degrees north and a fairly temperate climate. But I know temperatures hovering around freezing and damp air can be cold. I could envision having access to a real woods and a large supply of firewood; it was doable.

I also became disenchanted with making the drive with our car to the UD campus, only about 15 kilometers away, but at least a half an hour was needed to negotiate all the traffic lights and congestion. I rode a trolley, a bus with two electric pickup arms on its roof to supply electricity to a large electric motor. It was better than the diesel buses or automobiles, but that meant waiting for a bus and still I was in the traffic. An hour often was needed for the ride home or to school. So there were a few signs that living out might be better.

Joy was born in Winchester, Indiana, and her parents were farmers. They worked hard and I think competently. They milked cows when I met them and that in itself required commitment. But knowing Joy now for a long time, I think her parents considered providing material goods was the primary part of parenting. I cannot remember when my father-in-law said the weather was ideal, his crops were really good, and everything was fine. There always seemed to be something out of place, the weather was uncooperative, and the neighbor was a bastard. Joy told me some ugly things about how he treated his two children, Joy and Jimmy, and it is not worth going into that. But I know that Joy was left with the feeling that men were not good, except at that time me. She really needed to get away from her father and maybe her mother. But Joy was sympathetic toward Afro-Americans and Indians, and she set me straight on that. My family had not confronted that problem, if it is a problem, and while they did not say anything derogatory about those groups of humanity, they actually had no experience around them, so it was an abstraction. Joy had a dear friend who was colored, if you will, and spoke only good of her. I believe her and know that Joy has it right. She grew up on a farm, and when she married me she ended up stuck in towns. And after about seven years of this she started to push me toward moving into the country. We were living in Kettering, Ohio, a large suburb of Dayton, and there were probably about 500,000 people in the surroundings, just lots of traffic and noise, and little of nature.

We looked here and there for a small farm to move to. We went as far as Hillsboro, Ohio, which had to be 80 kilometers south and east of Dayton looking for a place. So we would have been content just having a place over the hill, so to speak, so that we could urinate on the ground without someone seeing us. In the summer of 1972 we found our new home, a 26-acre farm on Ohio 185, in Darke County. We were about a year late, however. It had been a 40-acre farm on both sides of the road, the north part bordering the Stillwater River. So we had the south part and there were six houses plopped on the road on the north part. But, hey, we had a small quarter hectare of woods, a beautiful square barn, a hog house, a corn crib, a chicken house, a small yard shed, and an old brick house. The house bricks were fired on the property a hundred years before, according to the neighbors down the road. So as I walked the land I couldn't imagine how this all could be possible. It must be a dream. I was raised in a town of 5,303, Winchester, and had lived all the time in town, including the latest in Kettering. I had read Aldo Leopold's Sand County Almanac, and it all fit in now. He wrote that when you are outside looking around, the land is not yours, but it all is. The neighbor may say he or she owned this or that parcel, but when you are there, I saw a woods, probably three hectares to the south and east, and it was, at that instant, mine. So it turns out that the Indians were right on in the truth: We don't own anything, but then again we own it all together. Somehow Christianity and the Bible neglected to convey this truth to people, and it has become a disaster through at least the past 2,000 years.

So on September 2, 1972, we moved to our new home, our farm in Bradford. Our house in Kettering we sold and were able to realize some equity so we were not destitute financially. But it surely changed our lifestyle. I was ready for it, even though I was still doing graduate studies with teaching. I was comfortable with it all, and my first acquisition was a tractor. Joy actually found an advertisement for a 1948 John Deere B in Palestine, Ohio. I called them and the man was apologetic; it was really a 1938 John Deere B, not a 1948. That seemed even better to me, and I jumped at the chance to get a tractor, a real tractor. The tractor was called unstyled, but it seemed to have more style than any styled tractor. It was about 20 horsepower, with a drawbar pull and flywheel start, and it ran on kerosene and started with gas. It was a beauty in my eyes, and I hurriedly was north of Palestine to get the machine. I drove it home, killing the engine once about three kilometers just outside Palestine because I still could not get the idea that the gas was a lever on the steering column, the clutch was a long lever on the right side, and the brakes were two pedals, one for each rear drive wheel. I agonized quite a while getting it started by turning the flywheel to crank start it, but finally it went and I got home. I remember it being an overcast day in October, cool, but not snow yet.

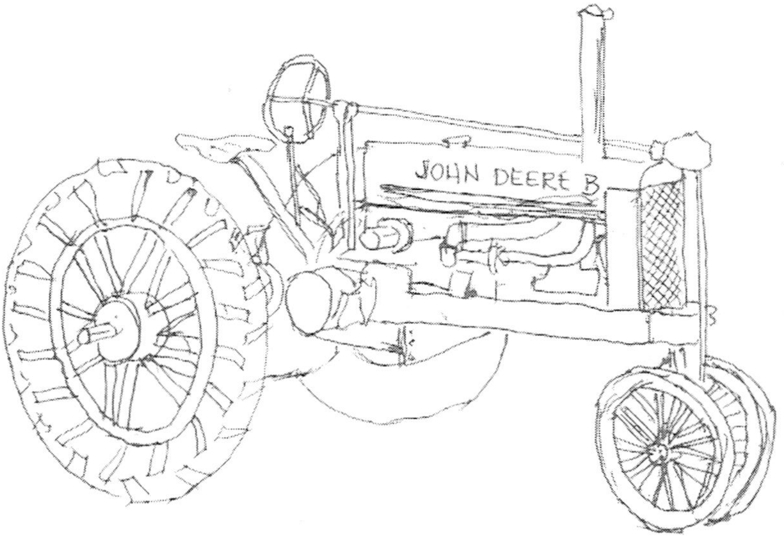

I always sensed that my father-in-law was disappointed that I had purchased a John Deere and an antique yet. He was a Farmall, International Harvester, man. Red tractors were better than the green ones, but he never came out and told me I had made a mistake. Well, the John Deere B was a sweet thing, almost alive. It even reminded me of a steam engine with two cylinders and chuffing and puffing. So it kind of gave me solace in knowing that now technology was not all glum. The tractor would serve me well. And looking back, I bought the tractor at a time when everyone was going big, big, big. They still are, but then antique tractors were just kind of a novelty, but in the course now of the last years, antique farm equipment values have grown more than the stock market. I got all my $350.00 for the John Deere tractor and then some. I wish I had it back.

I got a four-wheel box wagon from my father-in-law, a used wagon he had as an extra. I know he wanted his son to take over the farming, but that did not really occur. I know that my father-in-law was pleased to see my interest in farming, even if it was playing by his standards. He gave me a pull-type disk that turned out to be worn out, and during its first use, I pulled my back out and was laid up for three days. The disk was solid and heavy, but all the bolts were worn out, and so I ended up replacing most of the bolts, and it ran like a charm. I painted it green, and it served me well in Canada. My father-in-law found me a two-bottom, 12-inch blade John Deere Plow which I used, even in Canada. It was like new, having been bought new just as everyone was going to three, four, five and six bottom plows, so it quickly became useless for the serious farmers. I was satisfied with the plow even though it was a manual, requiring pulling a rope to trip the plow down and up. Too much hassle for the new farmers. And he found me a set of cultivators for the John Deere B, a two rower. I spend the better part of a day and a half figuring out how it mounted to the tractor, but it all went, finally. So I was set for farming. I had read lots of material on farming and agriculture, most of which today would be considered useless. But a real person wanting to practice agriculture (not agribusiness) should first read Wendell Berry's The Unsettling of America. Then Plowman's Folly by Ed Faulkner, and Rachael Carson's *Silent Spring* are musts. There are lots of other good reads out there, all of which the chemical companies and probably even the John Deeres and business would not want you to read. It all is really a clear choice for you; believe the business, war line or believe the truth.

Everything finally does develop or, as the biologist would say, it evolves. That is not a bad word, *evolve*. It means that things are moving in time. Since time cannot go backwards, it is going forward, evolving and not revolving. But some people want to shorten the thinking out of the various experiences, so they get apes and monkeys and whatever else tangled up in the word and it becomes a sad commentary on people, mostly religious types. One needs to forgive them for they know not what they are doing or saying, but that too is another story.

So in our evolution on our farm in Bradford we met up with some interesting neighbors. A neighbor some five kilometers west, Norman Giboney, was a kind of semi-retired tinkerer farmer. Norman only had 8 hectares maybe, a small farm, but he had a big Minneapolis-Moline tractor and a whole selection of equipment including a sawmill and a thrashing machine. He was old school and just played around. I think he was getting a disability check each month from the state, but I never asked about it all. He was rotund and jovial, probably in his late fifties at the time, and he did have a wife, Ruth, who seemed to be rather sickly. Norman told us about the Budget, an Amish newspaper that was published in Millerburg, Ohio, in Holmes County. Norman always seemed to be interested in the Amish and their lifestyle. I think he too was an anachronism.

And during the first year that we lived in the Bradford, Ohio, area our three oldest children went to Bradford School. There our two oldest, Kurt and Lori, became friends with some "Amish." Well, they were not Amish as we found out but of a sect called the Old Order German Baptists. But they dressed much like Amish, with plain dark-colored clothes; the girls had head covers and long dresses. We became friends with them and attended the church services. I remember the first service we attended, by invitation through our daughter Lori from the Diehl twins, Donna and Doris, was a foot-washing and communion service in the barn of one of the preachers, Joe Lavey. We were allowed to watch as the members washed each other's feet, the women washed other women's and the men other men's. They took this literally from a biblical passage. Then they had a full supper and communion wine and unleavened bread. We became hungry during the service, but it was interesting.

I think back now and write about this and see that Joy was a very busy person. While I was teaching during the weekdays at UD and going to graduate school in the evenings, she had to guide everyone to school, help them with their homework, make the meals, wash, and on and on. On top of it all, in April 1974 we had our fifth child, Charity April, born on April 14, nine days after Joy's birthday. Charity missed knowing her paternal grandmother, Oma, as Oma died on November 20, 1974, just seven months and a week after Charity was born. And as with all our children, Charity was a blessing beyond description. And she seemed to be, of all our children, the most calm and accepting. Not one to cry or fuss, but just be there smiling. Charity looked and I think acted a lot like her paternal Oma.

Eventually Joy was enchanted enough with the Old Orders that she actually joined as a full-fledged, baptized member. In the spring of 1975 she was dunked totally under in a nearby creek, Painter Creek. That gave her the authority to participate in communions and foot washings and have input at yearly meetings. She was not allowed to d rive an automobile although she could ride in them. She was not supposed to have electricity or a telephone, but since I was not a member we could have a telephone and electricity. We did purchase a four-seat buggy and a driving horse so we could go to church as the members did. After maybe a year, Joy was told that she couldn't have a piano and was dressing too modern. I think there were some other issues, but when two of the church preachers came and confronted Joy and admonished her to straighten up, she gave up the church. She still had friends, but by 1976 she had had enough and we were back to the "English" folks although much of what the Old Orders did seemed to be on the right track. They did drive tractors for farming, which seemed to be inconsistent, and this led to some members having the small Ford and Fordson tractors which could travel probably 60 kmph (35 mph). Out on a road, even a local road, that can lead to serious accidents, and accidents did occur. I felt better about this all and felt right from the beginning when Joy wanted to join the Old Orders that she would not last; Joy was herself, and no one was going to tell her what to do or how to behave.

Anyway, the Budget was a newsy weekly paper that was really letters sent in by Amish and Mennonite people from all over the world. Many of the Old Orders subscribed to the Budget too. In the sorry history of humanity, the reformation has to hold a good accounting. It is nearly impossible to figure out the truth, who was right, or what positive came of it all. My own take is that Christianity went downhill from the time of Augustine's supposed conversion to Christianity and has been sorry ever since. The Holy Roman Catholic Church became powerful enough to dominate Europe for over 1,200 years. And that meant spreading misery to people in the name of Christ and to bring profit to the church bureaucrats. Well, some folks had enough of it all, and in the late 1200s and on the reformation started. It was nothing more than a religious war, and that must be the worst kind of war. So many people were slaughtered and left homeless that it defies descriptions or normal belief. I have a seriously hard time thinking that Christ was in any way associated with the churches. The protestant movement, including the Lutherans, the Methodists, the Calvinists, the Presbyterians and others, protested the Catholic policies and views on religion. In between the Protestants and the Catholics, thinking that the main point of being Christian is to love one another, including your enemy, were the Waldensians and many Anabaptist sects. These included the Amish, Mennonites and Hutterites. There are many others, including the Baptists. These in-between groups mostly migrated from Europe, where the really nasty wars were being conducted, to America, Canada, and other more accepting places in the world. The Amish settled originally in eastern Pennsylvania and from there spread west into western Pennsylvania, Ohio, Indiana, and Illinois. They have groups all over the world now but a large settlement is in Ohio, Holmes County, and that is where the Budget is published. We subscribed to the paper mostly out of curiosity. And that is where we first read about a group of Mennonites living in the Peace River Country of Alberta and British Columbia.

The author of the letters from the Peace River Country was Carol Peachey, a young daughter of a Mennonite preacher, Jesse Peachey, and she described the various seasons in the area around Baytree, Alberta, which was about 50 km east of Dawson Creek, British Columbia. Baytree was about 30 km east of the British Columbia/Alberta provincial border. She was teaching the children at the Baytree Mennonite school and seemed to be a sensitive and observant person, one who appreciated all the various weather and Peace River country. Before reading her letters I had envisioned Canada above maybe the 50th latitude as a barren, cold and stunted-tree region. I could not envision verdant pastures and hay fields or barley and oats fields shimmering in the sunny plains. And wildflowers and plenty of wildlife, deer, elk, Moose, and bear. Even coyotes and wolves. And Carol described, over about two years of our reading her letters, the changing season from the lush summer with long daylight hours and haying and grain harvesting to the fast changing autumn and winter. But her description of winter was even attractive, and while 50-below temperatures seemed extra cold, we could do it all. So in my mind's eye of Canada I saw that we could go north, farm, raise our children in a pristine land, and get by, endure.

The Peace River Country of Alberta and British Columbia is tucked away east of the Canadian Rockies and below the tree lines and permafrost of the arctic region. It lies from about 55 degrees north latitude to up to 58 degrees, irregularly. It is country with the Peace River cutting across it from west to east, its headwaters in the high Canadian Rockies. The region is probably 500 km long from east to west with the Slave Lake an approximate eastern boundary and the Rocky Mountains the western boundary. It is rather a high uniform plateau with elevations from 3,000 top 4,000 km; it is a northern prairie. There are a few high hills that do not accommodate convenient farming, but most is slowly rolling with plenty of creeks and rivers. It was found by Europeans in the late 1700s, but the native peoples knew of it much earlier. I think one could argue that this area has a microclimate suited for northern farming, not corn maize and tomatoes and squash, but oats, barley and forage crops. Buffalo roamed here in times past and could in the future. *Blessed are the meek, for they shall inherit the earth.* It was land that would see long daylight hours in the summer and short daylight in the winter. The right combination of good ground and nutrients, enough moisture, and warm nights would provide microclimates for raising plants not expected at these latitudes and elevation. I was to find out years later that another 5 degrees north latitude above the Peace River Country, the area around Fort Vermillion along the Peace River as it meandered north here in a broad valley, was a microclimate that one was able to raise maize and tomatoes, not as tasty as in the south but they did get ripe on the stalks and the vines. But we were focusing in on the Peace River Country around Baytree, Alberta.

Joy would write Carol and receive replies from her. We found out the details of her family and were invited to visit with them. So, during the summer of 1976 we rented a camper trailer and pulled it behind our station wagon, the 4,000 kilometers to Baytree. We had John and Karen Mitchell, members of the Old Order Church, stay in the house and look after our animals and garden while we would be gone for almost two weeks. John and Karen were recently married, and this was a good opportunity for them to practice homemaking.

We departed Bradford in late May and arrived in Baytree in early June. During the journey north I remember it being long. But we did have a camper where the children could ride during the day, and we had some flexibility with campsites or motel availabilities. Since we now had five children, one being little Charity only two years old, it was a vehicle load. And I remember about the third day out, we had gone through Indiana, Illinois, Wisconsin, Minnesota, North Dakota, and a large part of Saskatchewan. We were doing good on traveling, making maybe 800 km each day, and that would mean a five-day trip. Five days out and five days back would give us two or three days at the Peacheys' in Baytree, enough to scout out the terrain and neighborhoods. We had driven a good stretch, from the North Dakota border with Saskatchewan to past Saskatoon. We were pulling the camper, so we were not breaking speed limits, but past Saskatoon we thought we could make a good run as long as daylight held up. So we drove, and drove, and drove. It was still daylight and coming unto 11 p.m. or after. At Battleford we decided we had had enough, so we rented a motel room at around midnight, with still dusky skies, not dark. We could have driven all night, but this then was our first encounter with the northern summer daylight. In the lower latitudes, 45 degrees down to 35 degrees, there is a bit of a change, but when one gets up to 50 or 55 degrees north latitude, it really is pronounced.

We visited with the Peacheys for maybe three days, looking the country over and seeing if it would suit. Our children were all enthusiastic about it all, and Kurt found a new friend, John Peachey, who was one of the Peacheys' sons and about his own age. Carol Peachey was older than all of our children, and they also had an older son, David. A younger son, Glen, kind of warmed up to Tim. And Jess and Faye Peachey, the parents, were very welcoming and encouraging. They surely must have sensed that we were serious about moving to the Peace River Country but also skeptical about whether or not we could do it. We did need Canadian sponsors, Canadian citizens who would recommend us and also officially underwrite the effort. With such a sponsor we could then apply for Landed Immigrant status and move to Canada permanently. The Peacheys were willing to be our sponsor, and that turned out to be a major hurdle that we had overcome by visiting with the Peacheys and having them vouch for us, so to speak.

During our few days in the Peace River Country we were treated to a snow shower on about the second day. It had been cool during our stay, and it was mostly gray overcast weather; not the welcoming weather like one would expect going to, say, Hawaii. And we all seemed to not be discouraged with this country. The Peacheys told us that there are an average of 100 frost-free days each year but that the beginning and end were always in question. So this year of our visit, the first week of June was not frost-free, but maybe the next three weeks of June, the four plus weeks of July, the four plus weeks of August and a couple of weeks in September would be frost free. The day when we left was warm and sunny, and that left us with a good memory of what we could hope for in our new home.

When we arrived back in our farm home, every one of us seemed to be ready to move, but now we had to do some real preparations, including getting Landed Immigration papers from the Canadian government, deciding how we would sell the farm and anything that we didn't want to take with us, and get immunization shots so we wouldn't carry anything bad to Canada. It would be a busy time....

CHAPTER TWO
Preparations for the Move

We had gone through much searching and thinking about what to do, and for Joy and me it always seemed to come around to going ahead and moving to Canada. Our visit to the Peace River Country was interesting and gave us hope. We knew that we would need probably two years to complete all the necessary transactions and preparations before we could realistically be moving.

And we would have quite a few distractions and detours to get it all done. I was working on my dissertation for a doctor of philosophy (PhD) degree in mechanical engineering and also teaching full time at UD. My dissertation work was funded through a proposal I had written, with help from my advisor, to the Air Force. So even here at the end I was compromising my principles in some way. But the work was going well, and it was interesting and challenging. I look back on it and realize it really was a good learning experience and there were some very important results, results that will probably go unread and forgotten in a library. It has been suggested that every technical paper or dissertation gets read completely by an average of one reader including the author. The key word here is *completely*, because I know that the members of my committee, faculty who were reviewing my work and measuring its appropriateness and quality, read portions of the document but not all. So it is really a testimony that humans have been successful at only three things: breeding, fouling their beds (the environment), and self-deception. If we would read just 1 percent of the documents that studious persons have written, we might learn something, but if we read the title and the preface maybe, we seem to know the important part of that book, or paper, or chronicle and that makes us learned and knowing. And we can then vote too. But I did go over a few hard spots in the work. I was writing the dissertation with a typewriter, a mechanical typewriter, and this meant editing could make me redo the complete document, if, say, a sentence or a paragraph was added early in the text. Being a technical effort, it could also mean the addition of an equation or figure, a chart, or a table would make a retyping necessary. And I had five respected faculty with my advisor on my committee. They all were very competent and caring but also rigorous, and twice they suggested major revisions to the document after I was well along with it. The first time my advisor managed to convince them it was good as it stood, but the second time I was faced with a major retyping. This sounds trivial now with word processors, but then with a typewriter we still numbered pages and the final document needed to be eraser-free, no smudges, and with good format. Fortunately the right margin did not need to be straight, but the left did. I came within a telephone call of quitting, calling a halt to the whole effort. But somehow my endurance and Joy's encouragement carried me through. I would receive my PhD degree in July of 1978, just a month before we moved to Canada and exactly 10 years after I received the Master of Science in Engineering Science from UD. That was awarded while we lived in Kettering. And at the time I received the PhD many people I knew were just astounded that I would do such a thing as move to Canada; why not earn lots of money now, what would I do, and so on. But they were the ones who would never understand my answers, so I just tried to endure and forgive.

In 1975 I believe, Joy was going to her doctor since she was still nursing Charity, and he recommended that she not have any more children. The birthing of Charity had done some damage to her, and the simple way to avoid any complications was to make her unable to conceive again. Well, Joy and I both were not sure of the health system in the USA either, and it seemed that always the simple way was the best, and it probably made more profits for the system as well. The standard medical profession took a very long time, many years, before they accepted the fact that chiropractic medicine had some validity, and that was always with a begrudging and obstructionist attitude. The naturalistic, herbalistic, and holistic medicines had to be quacks and frauds. Homeopathy was taboo. Even osteopathic medicine seemed to be unacceptable to the medical industry. They didn't prescribe state-of-the-art drugs, except maybe laetrile and marijuana. They surely didn't put money in the drug company coffers nor in chemical companies or the standard businesses. They just had to be evil. But that was just another tremor or sign that the country was being controlled and run by big big business. We did not buy into all that, and we found homeopathic doctors who were practicing near Troy, Ohio. So, disenchanted with American business, medicine, and government, and the bulk of the American citizenry, we went to the doctor and found out that the first thing they wanted to know--there were a man and a woman--was our health history. They talked individually with each of us, including the children, and they mostly just listened and took lots of notes. They were big on knowing if we had been sick, what we ate, and how much sleep we got. Naps were good, I could sense, and I have since that time never apologized for a good nap 20 minutes to 2 hours. There was no probing or blood pressure tests or pulse rate testing. They did finely feel over the body and listened and observed. It certainly was different. And when we were done, there were no prescription drugs, no placebos, doled out or demands on what to do and what not to do. They did recommend eating vegetables and fruits, organic if possible, not too much meat, if any, whole grains and whole breads, and most of the things one would encounter in a health food store. The main purpose for our switching doctors was to see if Joy was indeed on the verge of physical collapse, unable to have more children. It was not what either of us needed, another child, since we had lots of other things keeping us going, but we just wanted to be assured that Joy was still healthy, or not. The doctors, both of them, said there was no problem with Joy that couldn't be cured with good food and healthy living. Surgery and drugs were not part of their stable, and they made it clear that they thought she was healthy. That was a relief, even if the verdict came from "quacks." We continued to visit those doctors until we moved to Canada, not needing to devour any pills, get radiation treatments, or pay outrageous bills. As it turned out, the medical insurance I had at UD went ahead and covered our visits to the doctors, a good redeeming sign that not all was lost here.

And wouldn't you know, during the spring of 1978, after we had our Canadian Landed Immigrant papers, Joy told me that she was pregnant with our sixth child. We were close during those years, and it just happened. We were not trying to prove anything, and it came about quite naturally, with no expressed intention to fill up the Earth. And as you, dear reader, might feel, we had a good pattern and I think good intentions in raising our children. We would have our sixth child in Canada.

When we returned from Canada, in 1976, we had the opportunity to purchase about 10 acres across the road and behind the newer houses. It was land that was part of our farm before we purchased the property, and it bordered on the Stillwater River. It had lots of possibilities but had been dug around on; someone wanted to make a pond in the middle, and that was a disaster, just a mud hole with no water in the cavity. And we also obtained another acre or so bordering our hay field on the west. So now we had around 44 acres. We would hold on to the 10 acres on the Stillwater River and not sell it. That would give us some options and a bit of a tie to Ohio.

During 1977 we made lots of plans and felt out the proper things to do. We would sell our farm, the 32 acres, at auction in July of 1978, but first we listed it in April, hoping to sell it before the auction. We would also dispose of at an auction much of the furniture, tools, and items we didn't need or want. We had some chickens, and they would be sold; we had a manure spreader I had restored, and that would go. My good friend Greg Williams came up from Kentucky and bought it. I remember that some of the locals wanted to buy just the spreader seat, as it was a "collector's item." Greg got the spreader and the seat.

And I was serious about raising draft horses, Belgians, and using them. That was another prod to go to Canada; I just didn't have enough time to give proper attention to my horses. My original horse was a mare, Bess, I purchased in 1974 from Monroe Miller. Bess was a good Belgian with all the traditional markings, blond with white mane and tail, white feet and unmarked except for a small "proud mark" on her right hind foot, probably a bruise while working or brushing up against a stall board. She was probably six years old and was owned by Monroe Miller who was an Amish and lived in Holmes County, Ohio. We had made connections with him through John Kolb who also was interested in draft horses and had a very similar train of thought. John and his family lived about 30 km south and west of our farm, near New Madison, Ohio, and he drove every day to Dayton to run his liquor and party store. It was really another case of a small world as his store was no more than a couple of kilometers away from our house in Kettering, on Wilmington Pike. So John wanted to raise and work horses, but he was in the same bind as I was. He seemed to be a bit better off financially than I was as he purchased a team of Belgians in Holmes County a couple of years before we met him. But he told us about Monroe Miller and that he probably had a good mare for me. I did want a mare, so then I could get foals and expand my stable. Bess came to me pregnant, and on June 17, 1975, she had a nice handsome stud colt. We named him Royal Jay, kind of using his daddy's name, Conell Jay, as a key. He seemed royal to us and was a beauty too, blond with white mane and tail and feet. We had the opportunity to sell him, and we did, on June 1, 1977, to Melvin Kuhns in Holmes County, an Amish. Then on July 9 Bess gave us another fine stud colt. We named him Darke Chief, and he was surely another good horse; he seemed to be a candidate for a chief. The Darke was the fact that we lived in Darke County Ohio. And that was interesting because it was named Darke County because it was a swamp land, with lots of dense forests when the white men came. Too bad it still isn't a swamp land, but now the farms need to be tiled to drain water or else it would revert to a swamp again...maybe, hmmm...

When we moved we would keep Chief with Joy's father who took care of him until, during the summer of 1979, we sold him. Joy's father was a horse admirer and probably one of those anachronisms as well. He was a serious farmer and couldn't see clear to use horses.

We also had a number of chickens, geese, some ducks from time to time, a few pigs, and a few cows and steers. Our all-time favorite pig, Petunia, was a family pet and had a personality to match. Looking back, I don't ever want to raise animals like we did then, as a commercial, "useful," utilitarian way. *Animal Farm* and *Charlotte's Web* are just two reasons we finally, as people, need to assess the way to interact with the other souls in the world if we are to become truly human beings. And we always had a dog or two. On January 16, 1976, our dear dog Girl, or Pooch, who adopted us in Kettering years before, died of cancer. We gave her a good recognition and burial. Tim, who was an avid reader and thoughtful soul, wrote an ode for her,

Ho Friendly, you mother's gone, up where everything's fun. We tried and tried to keep her, but we could hardly do it. Everyone was crying, while daddy was trying to put her out of pain. Finally I saw the spirit but I couldn't touch it, all I could hear her whispering was: bury her by Stillwater River.
Tim Rolle

She had a daughter, Friendly, who was a beautiful red Irish setter-like medium-sized dog we would take with us to Canada. And Friendly had a daughter, Prissy, who also came with us. We would leave Friendly in Canada. She would be run over by one of the local trucks on dusty Highway 49 just south of our home, and we would bury her on our new property, but that is getting way ahead of my story. But mostly I want to emphasize that we actually considered all of our animals as part of our family and it was going to be difficult to part with most of our farm animals.

And the weather was going to provide us with a preview of Canadian weather. Both Joy and I had lived in east central Indiana, and where we were in Ohio was the same weather. Winters were not grinding cold, and snow was problematic. The rivers and ponds and creeks never froze deeply, but winters could be cold and damp. It was not a moderate climate at all. So during the winter of 1976-77 it seemed to know that we were contemplating moving to Canada. Well, it would give us a taste of it. October and November turned cold, and December had some very harsh days of near zero Fahrenheit. Then in January we had a full blown blizzard with -20 C (-10 F) temperatures and a howling 80 kmph wind from the west. Our house was cold, and we had drifts probably 3 feet high. It was a chore just getting wood for the cook stove and the box heater in the family room. The cold weather moderated in early February, and that was welcomed. The following summer of 1977 was a dry one. And then the winter of 1977-78 became the premonition of Canadian winters, just not as long and dark. But again, October was cold and damp, November turned very cold for the season, and December was even more cold. January came in hard, and we had sub-zero temperatures, going to –25 degrees C (-16 degrees F) on January 10. And it remained cold so that in the evening of January 25 the wind began to howl the temperatures dropped again to below zero, and it started to snow. This was a full-blown blizzard, and it lasted through January 27. When it calmed down on January 28, we found our Ford station wagon completely covered with a snowdrift just east of the house. And there were 10-foot snowdrifts all over. The highway we lived on, State Highway 185, had over 3 meters of snow for about 200 km each way from our house. There was no traffic that would get through until February 1 when a large bulldozer came through and shoved the snow aside. Even then there ended up being 5-meter-high pikes of snow on each side of the road. I was unable to get to school in Dayton until February 2. We had enough food and wood and water, but we heard, and found out too, that the day after the blizzard hit, January 26, the local folks in Bradford had hoarded up all the essential food, bread, milk, butter, canned goods and whatever, so even if we could have managed to get to town, Bradford, the grocery store was empty. Another demonstration of people's ways. The snow would stay until late March, and it was March 5 when we dug the Ford station wagon out of the drift. So we were able to cope with all this, and it came in handy in Canada.

Our farming was continuing, and during the last two seasons in Bradford, the summer of 1977 and the summer of 1978, we had marginal drops. I had been check planting my corn for the past five or six years and found it a good way to avoid chemical herbicides, cultivating both ways. My check planter, an old converted horse-drawn John Deere two-row model, worked good. I always used open pollinating seed corn, Reed's Yellow Dent, Bloody Butcher, or Blue Clarrige. The names tell the story; Reed's was a traditional yellow corn, Bloody Butcher was red, and the Blue Clarrige was a deep blue, when ripe. All three of these corns were favored by my animals over the hybrids, and I often tried feeding some hybrid corn with my open pollinating varieties. The horses, the cows, the pigs especially, would let the hybrid lay and not eat it. But in the rush to maximize profits, taste is thrown out and so, I think, is nutrition. But the summers of 1977 and 1978 were dry in west central Ohio, and we even tried to replant some of the corn. Without rain, seeds do not sprout or grow, and it really was odd because of the sufficient and even excess amounts of snow during the winter that they did not translate into good even rains in the summer. We did get some hay, some oats, a bit of our corn, and even some soybeans. But it sure was not something to brag about.

But we were looking ahead and planning. During the last Christmas in Bradford the family made a special effort. I made the following entry in the log I started when we moved to the farm.

Dec. 24. 11:30 p.m. This evening we spent a very nice Christmas Eve. Charity, only 3-1/2 years old, enjoyed the Christmas more than her earlier ones. We had an early supper and after doing the evening chores, sat around the piano in the front room of our house and sang Christmas carols. Charity sat on the three-legged stool Kurt had made at Opa's for 4th (grade) and Heidi sat on the little bench that Kurt made in 8th grade woodshop at Bradford school. Tim read from Luke 2:1-35. Joy played the piano, and the rest of us sang. Charity sat and quietly observed all the events. Her eyes sparkled and enveloped all the shining lights from the tree and the lamps. Heidi also enjoyed it all. We opened all our presents, and everyone received more than expected-perhaps love too. The three older ones, Kurt, Lori, and Tim, had done much to prepare for this Christmas. Lori knitted everyone a warm pair of mittens which she designed also. Kurt stripped paint and finished an old rocking chair that had sat in the corn crib for two years or more. It was very well done and liked by all. Tim and Heidi had made craft items in school which were thoughtful, useful, and welcomed. I received so much more than expected that I didn't know how to respond. Everyone went to bed tired and content.

I read this back and know that I had left out many details, but I think it conveys the idea that we were one happy and loving family. We had much to be thankful for and much to look forward to. As I wrote before, we then experienced the January cold and blizzard, and that turned out to be a prelude for our ventures in Canada.

In the summer of 1977 we had applied for visas and Landed Immigrant papers for Canadian immigration. As with most of these sorts of things, we waited and waited for a reply, a response, and best, a date to make the legal arrangements. Toward the last of October, I think, we received notification that we needed to make an appointment with the Canadian consulate in Detroit, Michigan. This was promptly arranged, and we had an appointment in middle February to travel to Detroit and complete the documentation. It would be another step in our adventure.

I remember shortly after the blizzard of late January I was sitting in the back room watching a documentary program on the state of the world. The back room was an addition to the original brick house. It was 2 x 4 framed with insulation and probably about 6 meters wide by 20 meters long, all along the south side of the old house. It was actually two rooms, the west room being a small washroom for the washer and dryer and separated from the back, family room by a stairwell that went to the basement. The family room had a trestle table that we ended up taking with us to Canada, a rocking chair, and a small table with a black and white television. We often sat around the television to watch favorite programs. I remember it was late in the evening, the children and Joy were in bed, and somehow I became engrossed in a program dealing with the population problem, the energy problem, and the resource problem. It was a broad discussion of the crucial world problems, as perceived in our age. It brought out the fact that population is increasing exponentially so that if nothing happens to change the direction, we will become submerged in people. And they will demand comforts or at least resources and food to survive. It will all put a demand on the Earth that cannot be sustained. And the saviors of nuclear energy, the green revolutions of infinite supplies of food and unlimited mineral resources are an illusion. We probably are headed to first of all a heat death and, if not that, then a cold freezing death, humanity that is. It was all information and ideas I had read and considered before, and in the late hours of the day, at night while outside it was cold and my family was sleeping upstairs, I just saw everything closing in on us. I had the feeling that must be experienced by someone getting ready to jump out of the second or third story window of a burning building; do I jump or not? What was the hope or assurance that one would escape safely? Our hope was the upcoming move to Canada.

Our appointment with the Canadian consulate was very interesting. It was a cold February morning when we left. Kurt, Lori and Tim would hold down the fort. The drive to Detroit would take roughly three hours, and I believe our appointment was 10 or 10:30 in the morning. So we left at 6 a.m., probably arriving around 9 a.m. in Detroit. As we approached Detroit from the south we passed many derelict, empty buildings, warehouses and abandoned factories. It looked like the bombed-out cities of Germany, England and Japan must have looked like after the Second World War. It was surely a testimony to the economic, technical and social degradation that seemed to be occurring in the rust belt of America. It was just another reinforcement that we were doing the right thing in moving to the Peace River Country. We met a friendly, helpful gentleman in Detroit at the Canadian consulate. He had done some homework on our application and had some very pertinent, appropriate and probing questions for us. *Why did we want to go to Canada? How were we going to make a living in the Peace River Country? Are we sure of this move?* We gave him our honest answers and thoughts. We explained that we would obtain a quarter section of land, set up a small-time farming operation, live frugally and efficiently, and do some odd jobs if need be. We hoped that within two or three years we would realize profits or enough income to survive, and I might even do some consulting engineering work. He said the reason we would be accepted as landed immigrants was that I was an educated engineer who could find employment in that line. Besides, Canada could use more good engineers, implying that I was a good engineer. That I was a good engineer surely is an open idea. Anyway, we had our papers and were set to move. Another hurdle passed.

So now we were, if not absolutely committed, psychologically committed to moving to Canada. We were getting correspondences from the Peacheys fairly regularly and were moving ahead. It came as somewhat of a surprise when the Peacheys wrote in the spring of 1978 that when they planted their garden they would also plant some things for us: potatoes, carrots, turnips and onions. That was such a nice thing to even think about, let alone do. The plan was to move in late August, after my PhD work was finished and after we had disposed, if that is the right word, of those items that we didn't want to take with us. As I wrote earlier, we put our farm up for sale in April 1978 and planned an auction sale in late June 1978.

We had purchased a one-ton Chevrolet C 30 truck from Loren McMaken that had a stock rack on it. The roof of the rack was raised so Loren could transport his draft horses. Now he wanted a newer truck, so we had the perfect vehicle to move two horses. He also had a large two-axle horse trailer. It was designed rather oddly in that it was tongue heavy, the axles were placed toward the rear. As it turned out, when two or even one horse was in the trailer, their weight would tend to move the load on the wheels to the back and lighten the tongue load. It was a nice trailer with a compartment in front for feed and water and whatnot. Our dogs could ride there too. So we had now the perfect rig for our move; a one-ton truck with high racks and a trailer to move two draft horses, Bess and Morning Star. I remember Loren McMaken and his wife, Pauline, being good neighbors. They actually lived probably 15 km from us, but they came over to our farm and we visited them too. They gave us lots of pointers on having horses, handling them, raising them, and enjoying them. Heidi referred to them as "Pauline and McMaken" in her own inimitable way. It would be the main vehicle set for our move. And then in middle June, June 20, 1978, we purchased a 1969 Chevrolet half-ton truck. It had a camper shell and lots of extra lights, fog lights, lights on the top of the camper shell, and trailing lights. It had mud flaps and was a kind of citified truck, one that had lots of superfluous extras. But it had only 40,000 miles and seemed to be in good working order. The engine was not big, a small v-eight, and it was a rear-wheel drive with regular tires and would be struggling in the icy, snowy roads. As it turned out, the truck served us well for the time we were there, and it was finally transferred to Kurt, who transferred it to Tim later. But that is another story.

Our auction sale was held on June 24, 1978, and it was a typical auction sale. These sorts of events are quite a bit of social gathering and enduring and getting on with the show. We sold much of our equipment: two plows, a manure spreader (which I mentioned earlier that went to Kentucky), a corn binder and corn shredder, and lots of small items. We sold all the animals except the two horses, Bess and Morning Star. Amy was a young cow with her first heifer calf, Molly, our five-year-old milk cow, some chickens, and the hay. All in all we realized $3,807.80 with expenses of $579.97. This would surely help us for our move and settling in Canada. The 34 acres sold for $66,500. I remember discussing the sale of the farm with our auctioneer, Ralph Hoblitt, well before the auction. We had decided that we needed to get $70,000 or something thereabouts. It was probably the going price at that time, about $2,000 per acre, and that seemed fair. At the sale there were only a few--one never knows who is actually bidding--bidders for the farm, and it got stuck at $66,500. I remember sitting in the house trying to sort it all out; was this a sign that we should stay put here, or should we go ahead and take a lower offer? In the end we went ahead. There seemed to be an inertia that was now pushing us to sell and move. I also remember a week or so before the sale having lots of second and third thoughts. Was this really what we should do? In thinking back now I wonder if it was not an excuse, but I thought about Joy, who was a driving force in this adventure but who could flip around and change courses quickly. She was always with me on moving, which we had done four times in the first 16 years of our marriage. Often she was the driving force then too. But was she really one who could do such a drastic move that we were contemplating. We would not have a reliable paycheck or a steady income, and without electricity or telephone, could she continue? I really had my doubts, but looking back years later, maybe it was me who couldn't handle the stress and worries and hardships. Anyway, when I asked Joy if she still felt the move was right, she emphatically said *Yes*. So we continued. And I think we finally did get enough money for the farm and the auction sale that we could remain in the clear, without borrowing money from other people or some financial institution. We were determined to not go in debt in Canada. As Ben Franklin wrote, or said: *Borrowing money and going into debt is a form of slavery.*

Late in July we had Bess and Morning Star serviced at Monroe Miller again, the sire being Conel Jay. They would each have a colt in Canada, and then some. Our horses were productive and healthy and part of us. We drove to Monroe Millers on July 31 to pick them up in our one-ton truck. They rode well in the high stock rack, and I remember feeling that the move was going to go all right.

On August 4, a boxcar was spotted on a stub siding in Versailles, Ohio, next to the George Worsch Lumber Company. We had purchased many building supplies at George Worsch, and it seemed symbolic that we should load up a boxcar with our belongings there. The Boxcar, again a premonition or whatever, was a Soo Line boxcar, 50 foot steel, built in 1973. Probably built in Fond du Lac, Wisconsin, where many of Soo Line cars were built and repaired or serviced. The boxcar was Tuscan red, typical of many of the railroad's car colors. This looked beautiful and solid too. It had the number 18708 and would carry many of our belongings to Dawson Creek, British Columbia. I was a railroad fan, so this was a real experience for me. I never imagined that I would have the opportunity to work around a boxcar, let alone load it with my material items. I had been told a week before that the car would be delivered on August 5 by Conrail, the going railroad through Versailles at that time. Conrail was a two-year conglomeration of the remnants of the Pennsylvania Railroad and the New York Central Railroad that combined as PennCentral. That union only lasted for two years, and Conrail was the result after courtroom pondering abut bankruptcies, public service concerns (?), and so on. The railroads were regulated to such a degree that they could hardly survive. I think the trucking companies, the airlines, the auto and truck building industries, and even the oil companies were scheming how to completely do away with railroads. But in the last year of President Carter's administration the railroads became deregulated, and this started a revival of the railroads. That would occur in 1979, and the succeeding president would receive the credit for saving the railroad industry, even though he had nothing to do with it. He was an actor.

We had made lots of plans on what to take with us, and what was crucial for the first few days in Canada and what could travel by boxcar. I made layouts, footprints, of the major items we would load in the boxcar: the two tractors, the hay loader, the two wagons, the disk, the grain drill, and some other items. I have put them in the Appendix for reference. The farm equipment was loaded on the boxcar: the two John Deere tractors, the hay mower, the wheat binder, the seeder, the disc, the two wagons--a green one and a high-wheeled wooden wagon--a hay loader, the cook stove, two round oak stoves, the trestle table and benches, and many small items. We decided that Joy's piano would go in the big truck, sheltered. The beds, dressers, many of the chairs and one smaller table, clothing, and just lots of small items would go either in the boxcar or the trucks. We had the one-ton truck loaded to the top. It had leaf springs, and they were nearly flattened, so that the axle was nearly touching the frame. But it just seemed to want to go to Canada. The horse trailer would take the two horses and the two dogs and enough feed and water for the trip. The trailer also had some cubbyholes for stashing small items.

The half-ton pickup would pull our two-wheel trailer that we were given by Joy's dad when we started farming in Bradford. It was an old Terraplane front-end axle with a thick wooden floor. It had hauled many loads of grain for us to and from the Alswell grain elevator in Bradford and firewood from the woods. Now it would be asked to carry a load to Canada. And what a load it would be. We had side racks that I made for hauling cattle that served well, and on top we draped a tarpaulin to keep the load out of the weather: rain, snow ??, or whatever. And the trailer was full. The pickup had a camper shell, and we managed to nearly fill it up as well. We would travel with me driving the one-ton and Joy driving the half-ton. Both trucks had a bench seat, so two or three of the children would ride in the two cabs. If it got messy, someone could actually ride in the camper shell as there was some room there.

It took us five full days of packing and loading before we were ready. And without the help all those days by a dear neighbor, Gary Berning, and his son, Mike, and one of my students, Don McDonough, we would not have made it in time. I made cut-outs of the various large items I was putting into the boxcar, and those included the John Deere A and John Deere B, the box bed wagon, the flat bed green wagon, the mower, the grain drill, and the hay loader. So I arranged them in the boxcar, and it was all a tight fit. Around and on top and under the wagons we put many items. There was to be a railroad inspector coming on August 9 to see if everything was acceptable. At that time the doors would be locked and sealed and opened only when the car arrived in Dawson Creek. Joy's parents also were helping in all sorts of ways. It was surely a reluctant activity for them; they must have not wanted us to leave, and neither did my father, but they helped as much as they could and more than was expected. Anyone who has ever experienced a move knows that many hands make lighter work, but the work is still heavy.

By August 9, 1978, we had completed preparations for traveling. I close with a short passage that I wrote the evening of August 9 in our home.

August 9. This evening the house is cleared out of furniture. The two-wheel trailer is filled with beds and dressers, the one-ton truck is nearly loaded, and the boxcar is cleared to go to Canada. The weather has been raining today after two warm sunny days, and the old farm is lush. The apple trees are filled as are the three pear trees. The weather has been poor for haying but good for growing. Tomorrow we leave for sure, and it will be done with mixed feelings. We hope it is for the best and that things will be good in Canada.

We slept the last evening on mattresses on the floor of our old brick house. In the morning we ate a kind of brunch under the big stretching mulberry tree. It was an old tree that was near the house and surely must have been planted and left to mature because mulberries are the preferred berry and fruit of many birds. It was my favorite tree on the farm and was just a very noble specimen. The birds would then leave the cherries and apples for us. There would not be apple trees or Mulberry trees in the Peace River Country, but there would be other flora for us.

CHAPTER THREE
The Journey to the Border

Leaving our homestead just after noon gave us a good half day to travel. With everything loaded down to capacity, and maybe then some, we had the one-ton truck with its rack loaded, and it also pulled the two horse trailer with Bess and Morning Star. The two dogs traveled in the front of the horse trailer, mostly. I drove this vehicle and mostly traveled at 45 or 50 miles per hour. The truck was capable of doing maybe 60 mph but I settled on 45 mph as a good speed; we wanted to get to our new home in Canada without trying for speed records.

The Chevrolet pickup truck followed. We had a camper shell on the truck, filled up with all sorts of items but also enough room so some of the children could, if they wanted, ride in the shell. The two-wheel trailer was pulled by the pickup. And it too was loaded to the top. I had made racks that extended the sides up to about 2 meters high, and we had a tarpoline covering the trailer load. Everything was strapped down, tied, and somehow arranged for the long journey. We sure were an odd caravan, *nicht vernünftig* (not reasonable). But we did have a goal and were on our way. It was a nice warm day, August 10, 1978, and sunny too. So on we went crossing the state line into Indiana at Union City and continuing on Indiana Highway 28. I tried to avoid the Interstate highways whenever I could, and even busy state and national highways were only used because they did seem to be mostly direct and probably reasonably maintained. The idea was to avoid traffic as much as possible without getting on obscure paths.

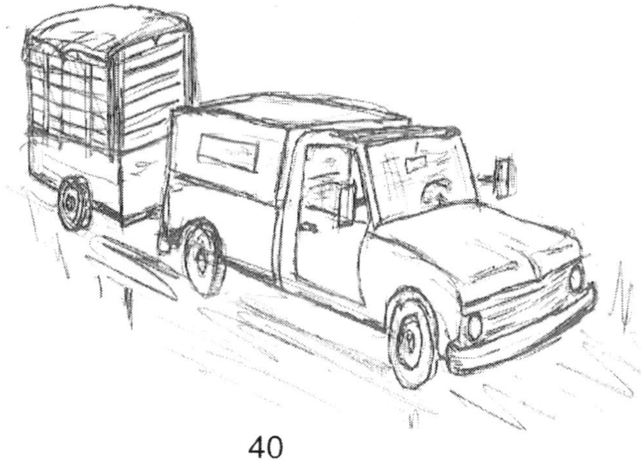

We had all the equipment and supplies we thought we needed. The big stuff was on the railroad boxcar, and we would meet that in Dawson Creek. The one-ton truck was not fuel-efficient, getting maybe 15 miles to a gallon, and its tank was probably a 16-gallon one. So every 200 miles we needed to refuel the truck. The pick up was maybe a bit more economical, but usually we just filled up both trucks when we stopped. Most gas stations did the pumping for us, as self serve was not yet common. But we always tried to make sure the radiator coolant levels were full, the oil full, and just generally checked the engine, wheels and other items. As the one-ton truck was loaded so much that the rear leaf springs had bottomed out on the axle, I always checked this area, to make sure it was still there. That was another reason I did not push the truck more than I had already. We only stopped once in Indiana, and I think that was at Kentland or another town in western Indiana. This would have been about 180 miles out from Bradford, Ohio, and it took us until five o'clock to make that. So we still had a long way to go. We decided that a good first overnight stop would be Kankakee River State Park, about 55 miles northwest of Kentland, Indiana. The roads we took were US 52 to Kankakee and then a secondary road to the park. By 6:30 we were getting settled in the park. It was not large, but the camping area was clean and roomy. It was also not crowded, although shortly after we arrived another camper came. I wanted to keep the horses as comfortable as possible, so I let them out of the trailer and tied them to trees. That way they could eat some fresh grass, get some fresh air, and just relax. The male camper came over and told us that he didn't like the smell of horses, and I was a bit surprised. But Joy had a quick response for him; she wanted to give him a clothespin to close his nose off. He turned and left, not bothering us the rest of the evening.

So we had a nice supper, the seven of us, with Joy about five months pregnant. The first day was uneventful, safe, and just a good start. Everyone I think was now getting comfortable with traveling and finally getting to experience some results of the anticipated venture.

It did rain a bit after we had supper, but we had our tent pitched and were getting relaxed. I believe we had three or four of us sleeping in the tent, and the other three or four slept in the pickup bed. We had mattresses for those sleeping in the pickup, to make it a bit softer. Those of us who slept in the tent had sleeping bags and blankets. We were nice and cosy. The two dogs, Prissy and Friendly, were good companions and not complainers at all. They were typical dogs.

The following day, August 11, 1978, which was a Friday, we packed up, put the horses back in the trailer, which they didn't really enjoy, I sensed, and put the dogs in the front of the trailer. And we were off to our next destination. As I look back on the day's travel and try to recall the highways we took, it is a bit dreamy. I have looked at the highway maps, and around Chicago it is very congested. We did travel two-lane highways when possible and avoided the interstates. That left us, I think, taking US 52 through Joliet, then taking Ill 59, which is now, in 2010, a four-lane highway. At that time it was two lanes and mostly rural going straight north. We then took US 12 through Madison and to Baraboo. At Baraboo we took Wisconsin State Highway 33.

I remember one close call as we traveled around Chicago. This was, for me, a major hurdle to overcome, getting around Chicago without any problems. I remember growing up and traveling around Chicago in the 1940s and 1950s. It was always a nerve-wracking experience because of the heavy traffic, and also the motorists were aggressive and non-forgiving. And it turned out it was similar for us, although mostly we made it without problems, except once. Highway 59 was a two laner and mostly followed the terrain of the land. Illinois in this area is mostly flat, so that was no problem. But at one junction there was a traffic light for 59 and an east-west highway. My truck was capable of 45 mph, but I needed distance to accelerate and stop, or decelerate. Traveling at 45 mph as we approached the light it was green for us, so I continued. As one could or should expect, about 100 meters from the light it changed to yellow, and I knew I had to continue because I would have been unable to bring the truck to a safe stop; I would have stopped in the intersection or beyond. So I just hoped that the traffic on the other highway, stopped and just itching to get going, would not jump the light. Well, I went through the light as it was turning red. Joy, driving the pickup, stopped at the light as I continued on. I still remember that incident and wonder why the cross traffic was not, in this case, taking off before the light was fully green for them. I was so relieved. I also took a mental note to try to anticipate changing traffic lights better, although in this instance I am not sure how I could have handled it. Maybe traveling at 35 mph would have done it, and as it turned out we did do that sometimes during our journey.

During my early years, when my parents and I traveled to northern Wisconsin--Couderay and Lemington--we mostly traveled Highway 12, US 12, around Madison. And that is how I visualized our trip. We traveled US 12 after leaving ILL 59 and around Madison. By mid afternoon we were near Baraboo, Wisconsin, and the scenery was getting real northerly, woodsy and not so congested. At Baraboo we took Wisconsin 33 west, which was a rough highway at that time. It was not a major highway but did take us through a number of towns, Reedsburg, LaValle, Wonowoc, Union Center, Hillsboro, Ontario, Cashton, and on into LaCrosse. This was all land that was unglaciated during the last two glacial periods, at least that is what the geologists and others say. The land was hilly and borderline mountainous. It was green with lots of woods, but in the valleys there were the typical Wisconsin dairy farms. In 1978 there were still many farmers milking, and I think the farms were not so large either. At Reedsburg we followed the Northwestern Railroad, and this continued to Union Center, where the Northwestern veered north. A branch line served Hillsboro, and I do remember a short train on that spur as we went through. Here with the pleasant and interesting scenery and the rather narrow two-lane highways, we were pushing the limits at 45 mph and I felt more comfortable. Hillsboro was a Czech community, although I am sure many Germans, Poles and other ethnic groups were here too. There was a banner across the highway entering the town that this was founded by Czech immigrants. A nice small Midwestern town. West of Hillsboro the highway, Wisconsin 33, became very narrow and hilly. It was up and down and then around. It was still very beautiful, but we were down to 35 and 40 mph. The truck labored up the hills, and I was cautious going down hills. The horse trailer had very good electric brakes, but after the incident on Highway 59, I was taking no chances. Out maybe 12 miles from Hillsboro we started to climb a serious hill. It was actually named Wildcat Mountain, and at the top was Wildcat Mountain State Park. It all reminded me of Kentucky and Tennessee in the Appalachian Mountains. The highway jackknifed back and forth and kept going up to a gap, at the top. Then we descended, and this was equally challenging, particularly with a more than 180 degree sharp right turn near the bottom. But we made it all together and passed through Ontario and on. In a long valley between Ontario and Cashton was an Amish community with their horses, austere dressing, and homes, and orderliness, *Ordnung*. At Cashton we crossed the railroad tracks for another branch line that served Viroqua from Sparta. I wondered how that line struggled up and down the hills of this area.

We arrived in La Crosse in the early evening and decided to stay in a motel. On the north side of La Crosse, in Onalaska, we found one that seemed good. We had to keep the horses in the trailer overnight, but we let them out for a short walk and also let the dogs stretch out. Everyone was doing good, and we seemed to have settled into a long quiet journey. It would be long but not always quiet, and that we would find out.

Staying in a motel was not as comfortable as a park, but it gave us all the opportunity to shower, eat a restaurant meal, and revisit some of the ways we were leaving behind. We also all had a good night's rest, I think. The horses and the dogs' having to stay in the trailer made me feel negligent, but we surely couldn't keep them in the motel rooms. We had two rooms, and as I recall, everything was comfortable.

The next morning was sunny and warm, Saturday morning, and we had a long way to travel. So we left Wisconsin behind, crossing the Mississippi River on the Interstate 90 bridge north of La Crosse. And we exited the interstate, following US 61 along the river for an hour or so, before we veered northwest on a state road, number 60, from Wabasha. I was trying to make a direct path without getting on the busy highways, and that became a real challenge. But Highway 60 was a good smooth road, asphalt if I remember it, and it was a pretty way. Minnesota here was just an extension of western Wisconsin with many neat farms and green woods. We drove an hour or so on this road before we then turned northwest on Highway 7 which kind of zigzagged north and then west and then north. We ended up after another hour or so on Highway 61 again, the busy U highway, but we didn't stay on this long as we then turned west on Highway 55 to skim under the twin cities suburbs. As I look at the road maps some 30 years later, this has been inundated with suburbia, but in 1978 it was still pastoral and nice. Almost directly south of Minneapolis we passed through Apple Valley, a small town on Highway 55. Shortly after that we joined with US 212 and headed mostly west. As we left the environs of the twin cities the surroundings started to take on the prairie look with more flat land and not many trees. It also was getting warm, although we did pass through a couple of local rain showers. About an hour west of the twin cities we again veered northwest on State Highway 22 which was headed in the exact direction we wanted to go. After about 20 miles we arrived in Hutchinson, Minnesota, and it was sunny and hot. I remember we passed the business district and a bank clock read 110^0 F. Don't hold me to that, but I know it was very hot. And we were all feeling the heat. As it turned out, I think this was also the reason we had some vehicle problems heading in our direction. So at Hutchinson we headed west on Highway 7. This was a state highway while the other Highway 7 was a county highway. It was getting into the late afternoon, and we wanted to make it to Montevideo as there was a campground, according to the map. It turned out to be a city park but a nice one, so we all settled in and had a nice supper. The horses were glad to get out of the trailer, and so were the two dogs. The weather had treated us well; we had a clear, warm evening for camping.

The next morning, Sunday, was clear and warm, and we packed things into the trucks and trailers and headed northwest on Highway 7 and joined with US 59. It was now flat prairie country and plenty of dry agriculture. Dairying seemed to be prevalent as well, and we also spotted from Montevideo on that there were many white seagulls, probably from the Red River Valley which here was flowing directly north to Manitoba and eventually into the Hudson Bay. But that was a long way off, and it seemed strange to see so many seagulls, but here they were. At 25 miles north of Montevideo we started to notice a scraping noise from the rear of the one-ton truck, and when we arrived in Appleton after another half mile or so, the noise was unavoidable. We stopped along the side of the road, and it was clear that the driver's side dual tires were hurting. Actually the outside wheel had separated and only the inside wheel was taking the load, which was probably on the order of five tons of load. We scouted around and found that Appleton had a city park like Montevideo, so we limped into that park. It was Sunday, so it would have been unexpected if an auto or truck mechanic was available or just a store for parts. And of course everything in town was closed.

We had up to this point experienced a safe and rather smooth trip, but this now was something else. The one-ton truck had dual wheels on the back, and the rims were what were called split rims. The rim had a ring on the outside that better facilitated tire removal or change, and the outside rim's ring had separated. As Kurt further inspected the situation, and he was beginning to show his gift of mechanics skills, he thought we would be best in getting four new or good used split rims as a prudent move. But we still didn't know where to get such rims. So it turned out that a local man gave us the local situation. Appleton had no auto parts store and only one mechanic who would be available on Monday. And the closest town with a possibility of having split rims was Willmar some 55 miles east of Appleton. The man also knew of a major salvage yard, agricultural, truck enterprise there, so we planned on journeying to Willmar the first thing Monday morning.

Meanwhile we just had to settle in for a day of rest and recuperation. Appleton was just a small town of maybe 1,000 people and quiet and sleepy. We had no problems here and spent the day in anticipation of the following day's work.

On Monday, bright and early, we drove the pickup truck to Willmar, Kurt, Tim and I. Joy and the girls stayed in Appleton and kept busy with cleaning and cooking. By this time we had accumulated some dirty clothes, so clothes washing was a task at hand. Out trip to Willmar, 55 miles out and 55 miles back took the morning away, but we did get four used split rims that seemed to be in good shape and were the correct size. Our tires were all good yet, as I had had them checked in Ohio before we left. By three o'clock in the afternoon we had the rims all replaced and were ready to go again.

So we left Appleton on Highway 7, heading almost due west now and then north. This was a very interesting part of the world. It was a kind of continental divide as the rivers and creeks south of here flowed south to the Gulf of Mexico while north of here the rivers and creeks headed north and flowed toward the Hudson Bay. To our left was the Big Stone National Wildlife Reserve, which helped solve part of the puzzle of where the seagulls were coming from or going to. It was beautiful country, and we were near South Dakota. We headed straight north on US 75 just east of South Dakota. The weather was overcast and had cooled down. It had the feel of rain, but that didn't materialize, yet.

After traveling about 100 miles from Appleton, thinking we now were back in the groove and making mileage and time, we suddenly again heard squeaking from the rear wheel of the one-ton truck, just the wheels we had fixed. So we pulled to side of the road. Here the highways had a wide shoulder, not paved, but smooth gravel and accommodating. The one-ton truck was leading and the pickup behind. With the one-ton pulling a large horse trailer, the pickup pulling a two-wheel trailer, everything packed to the top, it must have really been a sight. But we no sooner had pulled to the side of the road than a farmer and his wife and three children walked up from the east. We then saw that to the east was a farmstead. And it was comfort to have someone else be there to maybe help assess this new situation. The family was named Jorschaub although the spelling here may have degenerated. They were very friendly and helpful. Joy took up a pleasant conversation with the wife as the farmer helped us determine what to do. We, mostly my doing, had not tightened the lug bolts on the rear wheels enough, and they had worked their way loose over the past 100 miles. Actually only two or three lug bolts were still holding the wheels on, so again we had a major problem. But here it was, in the afternoon of a working Monday and we were only about five miles south of Wheaton, Minnesota. So we drove the pickup into town and at the Chevrolet dealer were able to get a set of new lug bolts, with the nuts. The lug bolts were serrated, so they needed to be pressed into the hub of the tires. Then the nuts were put on the bolts and tightened up. While we didn't have an arbor press to put the lug bolts into the hub, a sharp blow with a lead hammer, which somehow we did have, did the trick. Again, Kurt was becoming more and more experienced as we traveled, and so, after a warm farewell to the Jorschaubs, we headed north.

It was now late in the day, and rain was threatening. About 50 miles north of Wheaton we arrived at Breckenridge and then headed west, across the Red River and into Wahpeton, North Dakota. Here we ran into a real downpour, the rain seeming to have held off until our arrival. So we stayed overnight here in a motel as there were just no good campgrounds indicated on the map. We had planned to get much farther to a campground, but all our mishaps canceled those plans. But we had made some progress.

The next day was still rainy and dark, and we headed west and north. From Wahpeton we traveled Highway 13 west to Highway 18 and then north to Highway 46. We were jogging west, north, and west again to avoid the interstates and busy roads. We were now on Highway 46 headed straight west and traveling at probably an average speed of 35 mph. We took two hours to do 70 miles where we met US 281 north. Seventeen miles north on US 281 got us to Jamestown, North Dakota, and the crossover of Interstate Highway 94. Here we headed north and west on US 52, which would take us to the border with Canada. We could see progress now again, and patiently we headed towards Minot. The rain let up, and it started to get a bit sunny, although the hot weather seemed to have left us behind. We spent the day steadily heading northwest on US 52. We stopped for lunch, gas from time to time, and by late afternoon we arrived in Minot. We took another motel room, and here we let the horses and dogs out to stretch their legs. It was so encouraging to be with animals that were not complaining but accepting things as they came. And they really did appreciate the opportunity to walk around some. I always felt bad about doing the motels as then the horses and dogs had to stay in the trailer. But we would get more opportunities later. Here in Minot we were reaching the symbolic demarcation from the US into Canada, and that would occur the following day.

CHAPTER FOUR
The Canadian Journey

Tuesday, August 16, 1978, we departed Minot, North Dakota, in the morning. The motel had turned out OK. We again let the horses and dogs out of the trailer for a few minutes to stretch. Friendly, our dear red mixed lab and whatever was always accepting and stuck there. We called her Friendly because she was and also meant to please. Prissy, our other dog, too was always trying to please, and together they were good company. It was warm and sunny, so we all enjoyed the start of the day. Minot is about 100 miles south and east of Portal, North Dakota, which was at the border with and where we would enter Canada. The country around here is austere northern prairie with lots of brushy rolling land. The hills in fact can be high and steep, not quite mountains but surely not flat land. And interspersed with all this were rivers, creeks and washes which could be torrents of water during rainy seasons but mostly were just bottoms of valleys, dry and quiet. Cottonwood trees loved the valleys, and their roots surely sank in deeply and took in moisture, steadily or intermittently. But I will always remember Minot, North Dakota, for the fact that more weapons of mass destruction (in terms of destructive power) were buried in missile silos here than anyplace else in the world. And this was so ironic because a good friend of mine, who grew up with me in Winchester, Indiana, from 1941 until the late 1950s, went off to college and majored in business administration of all things and earned a PhD in that dark subject, went to Minot to teach at Minot State College. There were three reasons Bob Litschert went there: He liked the country and community, and he was a railroad fan like me and that was a major rail center with many Big Boy steam locomotives serviced there and which from that point west through western North Dakota, Montana, Idaho and Washington were used to pull long freight trains through the high Rocky and Cascade mountains. And Minot was in an obscure part of the world, well away from the hustle and bustle of the more populated regions. And Bob would have the opportunity then of seeing and studying the engines, the Big Boys which were articulated steam engines, single expansion of steam to two cylinders on each side, driving four large wheels for each cylinder and with two leading wheels, smaller than the drive wheels but big enough and two trailing wheels. They were identified as 4-8-8-4, and they surely were impressive. I remember seeing one at the Chicago railroad fair around 1948, and it was a big, big machine. But a big reason he chose Minot, away from everything, was that this was time of the Cold War, that contrived and brainwashing period, not unlike every other time in civilization's run to oblivion, where we had to protect ourselves against the evil menace of Russia and communism. I remember the Cold War times as a secondary issue for me. We were bombarded with lots of propaganda about how bad communism was, without explaining why, but of course we needed to have bomb shelters, drills at school for evacuation routines when the Bad Russians would bomb us and probably with nuclear weapons, learn how to crawl under our desks in case the evacuation process broke down, and stockpile food and water at home. The authorities never explained why we should do all this if a nuclear war commenced, since the world would be a cinder anyway, and when the food and water ran out, assuming we survived the initial blasts, what would we do? But the authorities know best and didn't want to explain it to the peasants, just like Hitler's Germany and Stalin's Russia. But we were "free." Actually I wondered about all this, wondering why we needed a bogey man to fight. But we had Russia, an "ally" to defeat the Germans and Japanese during the Second World War, who now was the enemy and whom we had to fight. So we built lots of "guided" missiles that had nuclear warheads. And we buried them in Minot, North Dakota, for the same reason my friend moved to Minot: to get away. And the irony was that the missiles were put at Minot while Bob was there, so he had, in a double irony, beat the War Department to the punch.... And now we too were trying to get away.

As a bit of a digression, my dear friend Bob lived three doors down from me on Richmond Street in Winchester, Indiana, during and after "the war," the Second World War. I remember growing up from about 1942 through 1945 bombarded with anti-German (Nazis, eh?) and Japanese (Japs, eh?) newspapers, magazines, neighbors and teachers. It just was all part of war mentality, but my family always told me that it was not the Germans but rather a political machine, the National Socialists with Hitler, that had started the war and that I should not be ashamed of being German. After all I am 100 percent German, by accident having been born in Wisconsin. At home my parents, aunts, uncles and grandparents spoke German often. So I picked it up and didn't understand why it was supposed to be bad. My friend, Bob Litschert, during the war often reminded everyone that they were Swiss. Actually his mother was Scottish, I believe, so he would have been 50 percent Swiss. But his name was surely German. And his father, who was the "Swiss," had a gun scope company right at his home. He had three men working for and with him making high quality gun scopes for the war effort. The government had given him a contract to make these items and the US government had also supplied him with machine tools, lathes, center-less grinders, drill presses and the like, to make the scopes. After the war was over he was given the machines and tried to sell gun scopes to "hunters and sportsmen."

In the 1970s when we visited with Bob in Blacksburg, Virginia, where he finally ended up teaching at VPI, we were talking about our families and I just asked him what part of Switzerland his family came from. His answer was a revelation, laden with lots of abstractions. "Oh, my family was German." So, dear reader, when I sound a bit contentious about our country, I think we all need to seek the truth, not the names that demagogues and just political parties want us to know. Do not resist evil, but love your enemy.

As we followed US 52 out of Minot, the Soo Line railroad tracks paralleled the highway for about 60 miles, and it was a pretty drive. The road and railroad were in the valley of the Des Lacs River, and there were lots of cottonwood trees along the river bank and up the hills to the east and west. It was now late summer, and there was just the hint of fall and winter in the air. The Soo Line would be the route taken by our boxcar, which too was a Soo Line. This was all so appropriate, as I implied earlier, in that my earliest loyalty to railroads was to the Soo Line, even though it was the trains and steam engines and passengers that developed that loyalty and not the corporation, who could care not a jot for the technology or the people.

At Kenmare the railroad tracks took a more direct route to Portal, and we followed US 52 which went straight north through Kenmare, Bowbells and Flaxton. It took us until noon to get to Portal, but we had no problems in that journey. So again, arriving at the immigration station we must have been an interesting sight for the people working there. The US officers had no direct effect on our leaving the US, but they did ask us if we had any valuables, drugs, weapons (of mass destruction maybe??), or other items that would be illegal. They just helped to assure us that getting out of the country was a wise move for us, with all the paranoia and grubby self-interests wanting to control citizens.

It turned out that Portal was a border town and straddled the US/Canada border and the Canadian immigration gate was just a few hundred yards down the road from the US gate. The community on the Canadian side was called North Portal, and between the two border towns there could not have been 500 people. It was a sleepy region but inviting too. There was a wide parking area for entering vehicles that a Canadian officer told us to park in. And then they too began their official scrutiny. It is always amusing and sad to see people like this who are just doing their jobs and doing it mostly mindlessly, trying to follow the rules and regulations explicitly. Much like gas chamber operators just wanting to do their jobs well, with no thought for moral reflections. But our paperwork was complete, our immigration papers that we had completed in Detroit way back in February were self-descriptive, and our passport was also acceptable. Since we had two dogs and two horses, that was cause for further scrutiny, and again our papers were in order; we had the horses immunized and the dogs too. No devious germs or whatever were allowed. As the officer reviewed all the documents and the vehicles and animals, and us too, I think he finally warmed up and realized that we were people too and maybe doing something he had dreamt of. But who knows.

So after about a two-hour delay at the border, we were now on our way through the middle Prairie Province of Canada, Saskatchewan. Saskatchewan is much like North Dakota, rolling hills and a brushy prairie. It is a beautiful country in its own way. And there were many farms here. Agriculture was very active and while the country was dry, with rainfall probably on the order of 30 centimeters per year (compared to Ohio where rainfall was typically 65 centimeters per year), the lower temperatures meant less evaporation of rain and moisture and more diffusion (soaking) of water into the ground. So the farmers had good ground and enough moisture to raise oats, wheat, barley and rapeseed (canola for the tarts). Corn and tomatos were not good options, but sunflowers were able to be raised. So every 10 to 30 kilometers there was a village or community with one or more grain elevators nestled up to a railroad siding. I remember the elevators were well kept up with nice paint covers, red with metal roofs and blazing letters "Saskatchewan Wheat Board" and with the large sheaf of wheat below the letters. Everything appeared nice and tidy. And it was a beautiful sunny day and warm. Traveling at no more than 70 kmph (45 mph) we could even enjoy some of the countryside. We traveled thorough towns such as Coalfields, Estevan, Hitchcock, Macoun, Midale, Halbrite, Ralph and Weyburn. Weyburn was only a bit more than a grain elevator and a general store, but on we went. The highway, number 39, angled northwest and by mid afternoon we had passed Yellow Grass,

Milestone and Corinne. We approached Moose Jaw by late afternoon, took a peripheral road around it and headed due north on Highway 2 for only about 15 kilometers where we turned again due northwest on Highway 42. We wanted to reach Douglas Provincial Park, situated on Lake Deifenbaker. The lake was named for John Deifenbaker who was the Canadian prime minister for a number of years. By early evening, maybe 7 p.m., we arrived at our destination and set up camp. Douglas was a nice park with lots of room, a washroom, and no other people camping. I remember that the terrain here was as we had experienced the whole day, rolling and dry. But here there were scrubby trees, not tall but coniferous and in due time could become tall. It seemed to be a park that was new, although as we would find out, in the north everything goes slow, except of course some people. The lake was the South Saskatchewan River, dammed up about 30 kilometers north of our campground and was a rather large lake, perhaps stretching 85 kilometers and wide enough that it was clearly a lake. Everyone of us was glad to be near the earth again, with tents and quiet surroundings. Even the horses were glad to get out, I know.

We spent a good night in Douglas Provincial Park and headed out early the next morning. It was overcast but not raining. It had rained some over the night but not too much, maybe 5 centimeters. And I remember the roads being damp. We headed north and west again. We selected highways that bypassed most of the larger cities, even though Saskatchewan was not over-populated in anyway. Regina, the Provincial capital, and Saskatoon were the only major cities, and at that time they each had no more than 35,000 people. But we still wanted to avoid them because of traffic. So we end up on Highway 4, a north-south highway that went to Battleford and North Battleford where we would meet up with the Yellowhead highway, Highway 16, which would take us to Edmonton, Alberta. But first we had to get to Battleford.

We came to Highway 4 south of Rosetown, and everything seemed to be going OK. Rosetown was a town of maybe 4,000 people, and there was an intersection of two highways, 4 and 7. Highway 7 headed east to Saskatoon and west to the Alberta border. We were at a relatively high elevation, maybe 650 m above sea level, and there actually were some ski resorts in the area. We didn't see them, but it was hilly now. Not more than 10 kilometers north of Rosetown we headed past an intersection with another highway, number 31, and noticed ahead that along the highway, which until now was paved and reasonably well-maintained, there was a sign indicating road construction. One could see ahead that the highway went down a long incline, maybe one kilometer down, at the bottom of which was a river. The highway had a high embankment which, 100 years before, would have had a bridge spanning the river, but now there was probably one or two or maybe three culverts under the embankment. The highway then went up again on the other side of the river and headed north. The road here was nothing but a graded surface, the pavement having ended before the descent to the river. It was dark and while not raining yet, it was not light. But nothing prepared us for what was to happen. I was leading and driving the one-ton with the horse trailer, as usual, and Joy was driving the pickup with the two-wheel trailer behind. As I left the pavement onto the grading it all at once became treacherous; it was like driving on a very slimy surface, which it was. It was our first experience with the gumbo of the northwoods, a clay material that was dusty and hard when dry but entirely different when wet. Since that time I have studied some about geological materials, and the terms are plastic and tenacious to describe wet clay. We know about that, and it is a good description. It makes for nearly impossible travel for a light vehicle with poor traction. While we were loaded down and had reasonable traction from my past experiences, this was even more treacherous than driving on icy roads, something I had plenty of experiences with in Indiana and Ohio. I probably went no more than 100 meters when I lost the truck; actually it wasn't lost but swerved to the right (or maybe the left?) and headed toward the edge of the embankment. My first reaction was to stop, which I was able to do. The trailer had not done anything out of the ordinary since I was not going more than 10 mph or 15 kmph. I was however about 45 percent with the grade and near the embankment's edge.

During all this, I had not observed that about halfway up the far grade was a road grader parked perpendicular to the roadway. When I was able to negotiate the truck and trailer to a stop, I saw that the grade began to head toward us. I thought this was a sheer accident that a big road grader would be here, but as it turned out, the highway department had this all thought out, knowing full well that people like me would be coming. So it took the grader a few minutes to come down the grade and up the near grade to us. It was big grader with three axles, and the tires all had chains, just as if it were plowing two meters of snow. It was probably a Caterpillar, but who was noting this? It was coming to help me out of this predicament.

It turned around, backed up to the front of my truck and chained solidly to the frame of the one-ton. Slowly we headed north on Highway 4 down to the bottom of the ravine and up the other side. I was literally being pulled through the clay, plastic and tenacious, and the road grader had enough power and traction to do the job. At the top of the far north side, the pavement commenced again, and after unchaining from my truck the grader headed back. Well, as it happened, Joy was driving the pickup truck and trailer and somehow made the same mistakes I made. She must have seen that I could not get through this mess without the road grader, but I am also sure she was determined to just follow in my path, hoping her abilities in driving would be enough to go on. After all, both of us had no experience with this sort of condition, and it just seemed unbelievable. Joy ended up getting even closer to the edge of the embankment, and the trailer even made her come almost around so she was headed mostly south more than north. I think back about this, and I know, from her and our daughter Lori who was riding with her, that it was probably the most traumatic experience she had up to that point in her life. I know too that the f act that she was more than six months pregnant made everything more traumatic. She was literally shaking and nearly shut off from reality when the grader came to her truck. It required more finagling to get the pickup truck turned around, but with the grader chained to the pickup it was finally accomplished, and Joy was back behind me on a solid paved highway.

Our vehicles and trailers and everyone were intact with no external damages. The trucks and trailers were covered with the mud, clay, plastic and tenacious, but we proceeded on. The road grader was indeed a bridge to our new adventures, and we wanted to thank him for his help, but he had other cars and trucks to pull through the muck. My, oh my, reader, it was true.

It was still early in the day, so we headed toward Battleford and then Lloydminster. It was uneventful, thankfully, for the 250 kilometers to Lloydminster, Saskatchewan and Alberta, slow but uneventful. We arrived in Lloydminster around 7 p.m., early evening, and we settled into a motel. Lloydminster straddled the Saskatchewan and Alberta provincial border, so we would now be in Alberta for the remainder of our journey. It was a good time to recoup our bearings and intentions. We had traveled for eight days now but were still a long way from home. But now in the north and about to stretch out more comfortably in the motel, we let the horses out and tied them securely to the trailer, feeding them some hay from Ohio and some oats too. The dogs were still happy and accepting.

The ninth day we had a good breakfast from our supplies and took the two trucks to a local car wash to get the most of the mud off them. The clay had dried some over the past day and overnight, but we managed to get them relatively clean. It was now on to Baytree, Alberta.

Well, we still had one more overnight, in Slave Lake. At Edmonton we had some tricky negotiation still. This was a large city, probably 250,000 people with all the irresponsible, in your face people on the highways. So we skirted Edmonton on the north, taking Highway 2 which is the main north-south highway through Alberta from Edmonton to Calgary and to the US border. But north of Edmonton it was a nice quiet highway, wide shoulders and little traffic. It was now clearly the Boreal Forest country with lots of poplar trees and also spruce and pine. But looking close one could see that much had been logged during the past century, so the old growth was mostly gone, replaced by the ubiquitous poplar. Eventually when people would become extinct, the spruce, pines and firs would blanket the countryside again.

Now we had another 250 kilometers to go to Slave Lake on the south shore of the Lesser Slave Lake but surely closer all the time to our destination. We arrived toward evening again. We were now at about 55 degrees north latitude, and that meant that daylight hours were long, even in late August. On September 21 we would have the Autumnal Equinox, and that meant the daylight hours are the same all over the earth: 12 hours. We were still in the long-day light so even at 7 p.m. there was some daylight left in the day. But it was now overcast and contemplating rain as we pulled into a campground about 10 kilometers west of Slave Lake. It was adequate for us, and the country was beautiful. We were, however, now ready to arrive.

The tenth day after leaving Ohio we departed the campground on the Lesser Slave Lake and headed west. We passed through High Prairie and came to an intersection with Highway 49. Highway 2 actually went north here to Peace River (the town) and circled back down to Grande Prairie. But that would have meant at least another 160 kilometers farther than turning left, south on Highway 49, going roughly 50 kilometers south and there heading west on Highway 43 at Valleyview. Valleyview is really just a wide spot in the road but is a major intersection of highways 43 and 49, and lots of trucks came through, mostly heading west on 43 to Grande Prairie. It was situated on high ground, and one could see for long distances. The prairie province or northern prairie could be typified by Valleyview. We would pass through this point quite a few times. Heading west on 43 it was now becoming much like a frontier setting. Well, frontier as the European would have it. The highway was paved, wide shoulders such that there could have been four highways placed in the clearing, but here land was so cheap that there was a wide clearing on each side of the road. I think back now, and even today, 2011, that one good reason for this was the fact that deer, elk, moose and even caribou were wandering around this country, and particularly at night they could wander (actually I apologize to the animals whose land this is, and they are just walking through it) onto the highway and get hit by a vehicle. It was not to protect the animals but protect the vehicles that the clearings were so wide. Anyway, it was still a long hike, 120 kilometers from Valleyview to Grande Prairie.

Six kilometers north of Grande Prairie we intersected with Highway 2 which here had wandered down again from Peace River, Grimshaw and other communities. We turned south, to the left on Highway 2, and went through Grande Prairie. In 1978 Grande Prairie had maybe 15,000 people and was rather quiet. There was an airport about five kilometers west of the city limits, the Alberta Northern Railroad passed through the town, and otherwise it was just there. Our boxcar would pass through, or maybe already had, Grande Prairie. Heading west we were now on the homestretch. We passed through Beaverlodge, and here we first saw the Rocky Mountains to the south and west. It was not a sunny day but clear enough to see the mountains, and even in late August some were snow-covered. Here in the northern latitudes that would be common, before global warming. Beaverlodge was a small city of 2,500 people, and it had a provincial agricultural experiment station here. I was given a book by Jess Peachey, The Peace Country Heritage written by E C. Stacey and based on the work and writings of Dr. W.D. Albright, the Dominion agricultural experimental station superintendent for many years, and who wrote about real agriculture in the Peace River Country, not the machine agribusiness that seems to have mostly overwhelmed all the world. The station at the time we passed through Beaverlodge looked just as it did in a photograph from the book. It doesn't exist now in 2011. Beaverlodge had two or three grain elevators, and being in Alberta they were painted white with the Alberta Wheat Board lettering. The Alberta Northern Railroad passed through here on its way to Dawson Creek. And we were now about 125 kilometers from home.

We passed through one more town, Hythe, before entering British Columbia and then on to Pouce Coupe. This was a small town, maybe 1,000 people just east and south of Dawson Creek, and from Pouce Coupe one was right at Dawson Creek. We could have made a short cut at Pouce Coupe to head to Baytree, but we didn't yet know of it, so we went directly through Dawson Creek. There were probably 5,000 people in Dawson Creek, and from here we headed back east to Baytree. Actually we crossed the British Columbia and Alberta border around 3 p.m. in the afternoon and the highway, number 49, was a dirt road again. In British Columbia it was paved, but here it was much like the graded part of Highway 4 in Saskatchewan, just not wet. It was dry and dusty. The Peacheys, who were where we were headed toward, lived about half a kilometer off the highway to the north maybe 15 kilometers into Alberta. Their farm was on an east-west road that looked to be the original highway approaching Dawson Creek from the east. It was now a passing road because of the "new" Highway 49 which was still being worked on. To the west of their home and barns was a one-room school, the old Baytree public school which was at that time the Baytree Mennonite School. Our children would attend this school for a little over one year. And we pulled into the Peachey's farm in the afternoon of the tenth day, tired but anxious and glad to have reached our destination.

CHAPTER FIVE
Taking Stock

We had arrived in the Peace River Country. It was the beginning of a new start that we had thought of for a long time, and the reality now was setting in. It was Saturday, and we knew that the first order of business was to get a place to live, even if for a short time. Of course, we intended to get a farmstead as soon as possible, but we needed to take it one step at a time. We stayed three days at Jess Peachey's, who was the main contact since we had journeyed to the Peace River Country two years before. The Peacheys, at that time, had six children: Carol, David, John, Julia, Twila and Glen.

Carol, the oldest, was our first contact in the Peace River Country when she wrote letters to the Budget newspaper. The Budget was an Amish and Mennonite newspaper published in a small town, Sugar Creek, Ohio. It was really just a series of letters from people all over the world who were most all Amish or Mennonites. The letters were newsy with personal items, and Carol was able to communicate the beauty and love she had for the Peace River Country. It attracted Joy and me to the area, and when we visited the Peacheys in 1976, they were very welcoming and encouraging. They were our sponsors for immigration to Canada. When we arrived in 1978 all the children were living at home except John, who was out in the "bush" logging. John was 16 years old and Kurt's age. But he took out on his own, seeming to not fit in with the restricted livestyles of the parents and Mennonite Church. During the three days we stayed at the Peacheys' we met numerous local people and the church members. With their help we found some available rentals, none of which were very fancy.

We finally settled on a rough dwelling on the Tomczak place. I think this was a homestead of a Ukrainian family by the name of Tomczak. They may have been Polish, but it certainly was a picturesque setting. It was on a quarter section square of land about two miles north and two miles east of Baytree, on the north side of a dirt road, and the present owner was a Gordon McRann, who lived maybe six kilometers north. He farmed the land for grain, mostly barley and oats. The land was flat, and the set of buildings was situated about in the middle. A creek meandered from south to northwest from the road. A plank bridge spanned the creek for the dirt road. And here in the Peace River

Country the creeks seemed to have eroded the banks so that the creek was four or five feet below the cultivated land grade. Poplar trees and a few black poplars, or Cottonwoods, grew along the creek bank, and there actually was about a hectare, maybe three acres, which was a small grove of trees adjoining the creek. A small Beaver dam had been constructed on the creek, and it appeared that a family of beavers lived there. A small beaver lodge was in the creek, but there was not a typical beaver pond.

As one turned into the lane from the road there was a hand-dug pond, the dugout, perhaps a half-acre rectangle lying on the right side. This would be the source of our drinking water until we moved to our own home, and it was edged by small trees and a few willow bushes. Some cattails and water lilies breezed in the water. As it turned out, the pond was rather shallow, perhaps no more than five feet deep at the center, and the water was full of squigglers, or short 2 cm-long worms. These squigglers became a source of irritation to some of the children as we used the water for drinking, cooking and washing. Boiling the water killed the little worms, but unless we actually ran the water through a cloth sieve or filter, they were still in the water. Often, when we had oatmeal or Red River Cereal one could see some nice pinkish short lengths in the cereal; they were squigglers that had been cooked. Our two oldest daughters, Lori and Heidi, were often involved in the baking, cooking and water handling, and they were very careful to filter the water from the dugout. We all appreciated their efforts but often teased them about leaving the squigglers in for protein and flavor. The neighbors and friend told us that in the deep winter of January and February the water from this dugout would turn black and take on a distinct odor of rotting algae. This provided us with another incentive to find a homestead as soon as possible.

But the home, for what is was, was an unpretentious single story building situated about 100 meters from the road on the left of the lane. It was essentially two small rectangular buildings, the oldest one actually a log building that faced the road. It had two rooms, a kitchen, perhaps 15 feet by 12 feet as one entered and then another room to the left, perhaps 15 feet by 15 feet. The second was added on the front and to the left so that together they formed a rough L-shaped building and was just a small bedroom, about 12 feet by 12 feet. The main entrance, entering into the kitchen, was on the front, facing the road in the older original section. There was another door to the left entering the newer addition. The building had been "remodeled" so to speak by putting a layer or two of red asphalt shingles on the outside walls. Many of the homes in this country were single story, but the decorative scheme was what we called a three-tone style. That was typically a light-colored band along the bottom, reaching up to the bottom of windows, another darker band of brown, green, blue or whatever, reaching up to the top of the windows and doors, and finally the upper band another light-colored band. We were not really impressed with the aesthetics of this, but it was not actually offensive and provided a bit of color during the long winter days. But our rental was a single red-shingle-covered home. The red shingles, however, were faded and worn. The roof was a black asphalt shingles, and it too was well worn. It did not leak, and that was a positive for the building. There was a basement, but it was full of water, as was typical in this country. Water just accumulated below ground level. And there was no electricity, telephone or running water. An outhouse was 25 meters or so behind the house and to the left. A few trees had grown up around the outhouse so that it was not conspicuous. Gordon McRann charged us $25 a month rent, and I think he probably was a bit embarrassed by that as he never insisted on collecting the rent. No one had lived in the house for a long time, and even the natives up here were probably wondering why anyone would want to live in the house. But we had committed to coming here and living as simply as possible, so this was really another extension of our camping experiences of moving up from Ohio. We knew we would not be here long, hopefully not more than a month. As it turned out, however, it would be a three-month stay in the Tomczak place. During our stay in these cramped quarters, we had been under a lot of stress and uncertainty for the past month or so, and it started to show itself. Joy and I had a number of ugly verbal fights here, and only the hope of something better kept the fights from festering, but maybe they did. I think back now, and I should have given her a lot more room, so to speak. She had many burdens right then, and she didn't need a contentious husband.

Directly behind the house about 40 meter back were the remnants of a log barn. I often went out to the area, imagining how it must have looked when it was new or at least functional. It was not a large structure, perhaps 15 meters by 12 meters, probably had a small hay loft and enough room for a cow or two and maybe one or two horses. The roof had deteriorated first, and the log walls were on their way. But up here it could take 30 years for a log to rot away because of the dry climate.

To the right of the lane east and north of the house was first of all a building that seemed to have been a chicken house in the front toward the road and at the rear a workshop for minor tools. A clump of poplars stood between the house and the chicken house. We strung a clothesline between some of the trees to dry clothes.

Behind the chicken house stood a rough badly deteriorating building that probably was used to shelter wagons or other equipment. Finally, behind that structure was another log building that looked to be a granary. This was probably the best preserved building on the farm, and I thought it was worth trying to move, if we found a parcel of land nearby. The logs were sound, and the roof too was substantial.

And our belongings were scattered around. We left much of the boxed-up items in their boxes and stacked the house bedrooms full. We did set up two sets of bunk beds and also the double bed. Charity was the youngest and slept on a cot. The kitchen would get the cook stove when the boxcar arrived in Dawson Creek, and while everything was very crude, there was space for the utensils, pots, pans, plates, and even towels and wash rags. Joy's piano, which we hauled in the rack of the one-ton truck, went into the big bedroom.

And our horses had a nice shelter in the log granary. I strung a very temporary fence from the road to the buildings so they could run around during the day. We gave them water from the dugout in a wash tub, and they didn't seem to mind the squigglers. The bulk of the equipment would just have to sit outside.

So we had a point for operations in the Tomzak place. Our four oldest children went to the Baytree School which was just to the west of Peacheys' home, maybe a quarter of a mile. It was called the Baytree School but was actually about a mile and a half west of Baytree. It had been a public one-room school until maybe 15 years before when the students were then sent to Bonanza. Bonanza was a small crossroads community six miles or so north and maybe three miles east of Baytree. A small schoolbus took the public school students there. But our students went to school at the old Baytree School, a Mennonite parochial school. We drove them there every day. I actually had some misgivings about the capability of the teacher, Carol Peachey, and afterwards other teachers, such as her sister, Julia Peachey, and later a David Donaldson. They all seemed more intent on obeying what the church preachers wanted them to teach than on trying to teach what the students should need in the way of self-learning skills, lifelong learning habits, and liberal thoughts, including forgiveness. But I did have faith that our children were independent enough to come up with their own thoughts, reasoning and results. I also had a bit of guilt, since all the people at this point, were very helpful to us and welcoming. As things went along, Carol did seem to be a good teacher.

We had packed most of the farm equipment and furniture in the Soo Line boxcar in Versailles, Ohio. It had not yet arrived when we did, but on the fifth day it came in. It was spotted on a siding in Dawson Creek right near the depot. We had checked every day for its arrival, and when it finally came, we were ready to unload it. We first needed to pay the freight bill from Versailles, Ohio, to Dawson Creek, B.C, and I had budgeted for around $2,500, which was the quote from the freight agent in Ohio. But when I went to the Provincial offices, that were located in the same building as the Dawson Creek post office, a bill for $5,000 was given me. That was a real shock, and I must have explained my predicament and misunderstanding that I had thought it would be half that. The authorities in charge were very friendly and accommodating and decided they indeed would only charge me $2,500. We were counting every penny and dollar, and this was very helpful to us. So we set about unloading the boxcar.

It didn't take as long to unload the car as it had in Ohio to load it up. The load had gotten a bit tangled around, but outside of a groove gouged in our dinner table, a trestle type table, everything was unharmed. And even the table was not too badly damaged, and I later filled in the gouge and it was like new. So the trip of our belongings was safe, if a bit slow. I can't even begin to imagine how much it would have cost us to hire maybe three semi trucks to haul our belongings, and things probably would have been damaged more. We had two wagons, a green multipurpose wagon and an old box bed that I had made a new set of sideboards for. We had the two tractors, the John Deere B and the John Deere A. So we loaded the two wagons, hitched one each to the two tractors, and drove them to the Tomczak place. That in itself took maybe three hours, since it was a 25-mile trip. But it all went well, and we were seeing real progress. We needed to make another three trips to gather all the items up, and for that we used the one-ton truck and the pickup with the green wagon and the two-wheel trailer. And without the help of Jess Peachey, his sons, John and David, and Jim Sadlier, the unloading would have progressed much slower.

Joy was now about six months pregnant and holding up exceptionally well. This was both her and my idea, and she displayed a real commitment to the venture. She could have just thrown it all up from all the ordeals we had had and could yet anticipate, but she didn't. The gathering of all of our belongings and having them at the Tomczak place seemed to be comforting to her.

Our big box bed wagon was very handy at this point as storage for much of the hand equipment and even some furniture. Even at that, much had to just stay outside in the elements, until we could find suitable shelter. We did bring along a chest freezer that had been loaded in the one-ton truck and that we put in a back room at Jim and Kathy Sadlier's. They and all the Mennonites had electricity and telephones. As we would find out, some of them even had record players, and one had a TV. These were not supposed to be owned by this Mennonite group, but people are finally people, and getting away with something seems to be part of our personalities and minds. Probably our souls too, if we really do have such things.

So we set up the cook stove in the kitchen, and I spent a lot of time getting firewood for it. The days now were nice and sunny and warm while the evenings cooled off, and nights were often cold. Not freezing yet but cold. It was now late August, so frost was a distinct possibility.

Many of the folks, the Peacheys and the Kandas, who lived just around the corner from us now and who visited us often, talked about the next "fullman" maybe bringing a frost or freeze, "eh?" I thought about that and finally realized they were saying the "full moon," which explained it. And of course the "eh" had to go into every sentence if possible. Then on September 18 we woke up to white frost covering the land. There was just the hint of ice on the horse watering tub. This was all signs of things to come.

During the summer the Peacheys had prepared a portion of their garden to share with us. They planted potatoes, carrots and turnips and this was such a good thing. They really were going out of their way to help us get adjusted to the Peace River Country. During the third week of September I brought Bess, our dependable mare, to the garden, and with the harness and a one-bottom plow we dug the potatoes. It worked very well, me leading Bess and Jesse Peachey holding and guiding the plow. We must have dug 15 bushels of potatoes for us and for the Peacheys. Jesse enjoyed doing this as it brought back memories of his childhood in Pennsylvania on his Amish parents' farm.

So we now had some winter food to store away. The trick up here was finding a place where the potatoes, carrots and turnips wouldn't freeze, so we squirreled the harvest away as best we could, knowing that we would be moving soon.

And one evening Gordon McRann's brother, Ken, came over and told us he had shot a yearling bear and wondered if we could use it. I was not a hunter and truly wanted to explain to him and everyone that this was not good practice, but there again I was being ungrateful. We went to Ken's farm about 10:30 in the evening of September 19 and took the bear carcass to Jim Sadlier's where we proceeded to skin and then hang the meat. The next day we finished cutting up the meat and put it away in one of Jim Sadlier's freezers, ready for some cold evenings this coming winter.

One of the early benefits of living here was the viewing of wildlife. On September 18 Charity and Joy saw two deer, a buck and a doe, from the kitchen window, grazing on the land. We would see many deer, moose, bear, and an elk or two. The bears were mostly black, but once in a while there would be a big brown bear.

It was flat and mostly cleared land, so one could see a good distance, and this all made for spectacular scenery. In the distant horizons though all was forests, bush as the natives would say, it was in its own way fetching and embracing. And wherever there were some nearby groves of trees there would likely be the ubiquitous poplar. Not a spectacular tree in size but surely one that adapted to the harsh winters, the dry seasons and the wet periods. They were the pioneer trees, so to speak, foresting the bare ground, when left to themselves and not cut or burnt. They always seemed so welcoming as well, straight and with a pale yellow bark, interrupted with horizontal marks at various vertical heights. It was indeed odd that most of the Canadians didn't seem to appreciate the beautiful poplars, or the sunrises or sunsets, or the timber or the wild animals, but many of them used that as an excuse for clearing timber from a parcel of land. I have often said that Canadians are just people too, no better than those in the States, but there were fewer of them, which made it better.

And slowly but surely the nights were getting longer, the days shorter, and the temperatures drifted downward, ever so slowly. There was not much rain although we did have a few showers. One day I went to Phil Sadlier's new home, which he was building on his land just maybe a mile north and east of our place. I spent much of the day helping him shingle the house, and we finished in the afternoon. Phil was the father of Jim Sadlier and Paul Sadlier, and he was retired from Montana, moving up to the Peace River Country to be near his boys. Phil was also a member of the Mennonite Church but took it all with some spices, so to speak. He had been raised Catholic, was disenchanted with that sect, and met some Mennonites in Montana. Their lure of preaching simple life and personal good works attracted Phil, as it did Joy and me. So we did have some good discussions with Phil, and he turned out to be the last church member we talked to before we left the church the following summer.

CHAPTER SIX
The Quarter Section at Pilsworth Corner

As we struggled to get our bearings in a new country, with a new climate, and with new neighbors we knew we needed to find piece of land to settle on. It all was taking a toll on Joy and my relationship. Mostly the children kept it all from getting too quarrelsome. In 1978 the land prices were mostly affordable. We had decided that a quarter section of land would be enough for us, both from a utilitarian and a financial standpoint. And repeating some of what I have already written, we had maybe $35,000 from the Bradford farm equity, the auction, and previous savings. We surely couldn't spend all our money and have nothing in our savings, and my farm equipment was such that about 60 acres of land would be pushing it, and me. So we set our target as a quarter section of land, preferably a section having some woods, trees, bush or timber, depending on who you talked with, and that we could get for somewhere between $15,000 and 20,000. We were near British Columbia, which was picturesque, somewhat more mountainous in the Peace River Country than Alberta, but also had land prices that were a bit higher, maybe 30 percent higher on average. And besides, our children were going to school at the Baytree Mennonite School which as about 15 kilometers east of British Columbia. So we began looking earnestly at land in Alberta, preferably near Baytree. There were many parcels of land, quarter sections of 169 acres, and they were all varieties. The trick would be to find the land not too far from Baytree, accessible from some road, and that did have some cleared land. The quarter section of land that we rented, the old Tomczak place, was nearly all cleared and was owned and being farmed by a neighbor, from whom we rented it. There was a nice stream running through it, and it looked like there was even a small beaver dam at one place in the stream. In addition there was a set of buildings, mostly in an advanced state of decay and a log barn that did not have a roof and was just in ruins. And the landlord did not really want to sell the land, so we struck that off.

Down the road from us about a mile east was a clear quarter section; i.e., no buildings but mostly land under cultivation. There was a small piece of woods, a bluff, if you will, of two or three acres, and it was nice, but again the owner was not interested in selling, except at his inflated asking price.

So we asked around. Most of the neighbors knew we were in the market to buy a piece of land, and they were exceptionally helpful in tracking down parcels that might suit us. We looked at land up long dead-end roads and even a quarter of a section or more past the end of the dead ends. The geography of the land near Baytree and up north to the Peace River, some 50 kilometers away, was mostly flat or very gently rolling. Highway 49 was a line of demarcation; north of 49 was the flat country up to and past the Peace River, but just south of 49 there was a very hilly region, all the way south to Beaverlodge and Grande Prairie. These were called the Saddle Hills, and the only way to get south was to drive 49 east to Rycroft and down Highway 2, or west to Dawson Creek and then take Highway 2 south. I thought that a parcel in the foothills of the Saddle Hills would have been nice since there was lots of woods including not just the ubiquitous poplar or aspen but some nice groves of spruce and pine. But this would have been rough farming, and that was something I had hoped to make the adventure go. So we journeyed every day around the region, traveling all the way east to Spirit River and Rycroft, north to the Peace River, northeast to Silver Lake, and west to the Alberta/British Columbia border. After a week of intense searching and not finding the piece of land we knew must be there just waiting for us, we were getting a bit discouraged.

Then we heard of a parcel at the corner of Highway 49 and Twelve Mile Road, called Pilsworth Corner, that was clear, had about 80 acres of woods and 80 acres of cleared tillable land. It was owned by George Blakus who lived up Twelve Mile Road maybe two miles north of Pilsworth Corner. We looked him up.

George was an archetypical Canadian, in my mind. He had a distinct British air but was indeed a true Canadian. He was a bit heavy set, maybe 1.8 m tall (5 ft 9 in) and had a rather deep voice. He was free with his "eh, eh" at any turn in the conversations and seemed to be a straightforward honest person. I think he was maybe 65 years old when we first met him. He had farmed here in the Peace River district all of his life, getting a compensation during the various wars because of his occupation: farming. He was of the old school of farmers who know just about when, based on the particular year and not the calendar, to do the crucial operations such as planting oats and barley and hay. When to cut hay and when it could be put up. He was one who earlier had put up loose hay and then the square bales and seemed to have the experience of it all. He burnt firewood for his small three-room house, which he and his wife Arlene spent lots of quality time in. They had one of the air-tight tin stoves, which I will talk about later when I talk about cutting firewood. George drove a faded blue half-ton Chevrolet pickup. It was a typical truck for the area; the windshield had a crack in it, one of the headlights was broken and it had numerous scratches. The Peace River Country was bad on vehicles, cars and trucks, but only because of the flying rocks and small stones on the road. They hadn't yet discovered salt as a means of controlling snow on the roads. I am not sure they are now doing it, because salt is ineffective in melting snow below about 10 degrees Fahrenheit (-12 C), This means that rust was only on vehicles that had wintered maybe 20 times. George's truck was creeping up on 20 winters but was probably only 15 years old. It had a camper shell covering the truck bed, very common up there and which we had too. George owned two quarter sections, I think. One of them was on the northwest corner of Pilsworth Corner, and he seemed to have had it quite some time. He never told me when he bought it or why he wanted to now part with it. But he was interested in selling it to us. I had the distinct feeling that he was just being a good neighbor and figured we would be the kind of people who would take care of the land. George always wanted to be sure that people took care of their land.

Arlene was from California, and as I understood it they met in California when George was on a "holiday." She too was a very pleasant person. We did find them at home when we drove over to see them. They were very cordial and interested in our family and were particularly interested in why we wanted to move to the Peace River Country. The word had surely gotten around about our moving in and must have been a very intriguing conversation item for the locals. I can just visualize them wondering why anyone from the "states" would want to settle in the far north with the cold long winters. And when they found out that I had an engineering degree and taught at a university, I am sure they really had a hard time understanding. That is not to say that we "understood," which we didn't fully, but we were there.

So we negotiated a price with George and Arlene, and it was $20,500. George probably offered $21,000, and we visualized getting just a tad less. Joy was a better negotiator than I was, but the offer was accepted. We had looked the land over carefully, and it seemed to be the ideal parcel. We spent the afternoon of September 18, 1978, in Dawson Creek at George's lawyer's getting the paperwork completed, and we were off. Since the land was in Alberta, it finally was recorded at Spirit River as a deed transfer. Finally, we owned the land with the legal jargon that it was originally, before George and Arlene, "Queen's land." And all mineral rights of Canadian land owned by small owners were owned by the Crown, or Queen. This is very common terminology and exceptions in all countries. It leads into the concept of what are mineral rights, and what actually does one own. When one thinks about it all, it is clear that we do not own anything, as the native Indians already knew going back thousands of years. And it really becomes circuitous, because is farming a form of mining that extracts minerals from the ground? Should the Crown get all or some of the crops because they are minerals? And what about the timber from the land? Isn't that too minerals? Finally, large timber companies, oil companies, mining companies, and any entity having enough financial backing and political friends can effectively circumvent the restriction of mineral rights; they get the minerals, they defile the land and move on. Oh, they do need to reimburse the Province and National government (the Queen) some percentage of the booty, but with good imaginative accounting and back scratching of authorities, they find it quite profitable. As with everything we experience in our local/state/provincial/national policies, the public, the small land owner or citizen, needs to just have a low profile, not make any waves, and go along with the body politic. Thinking too much or asking too many questions does not help anyone. And finally, as I read and believe sets it all in context, we are destined to just do things

Tätig zu Sein ist des Menschen erste Bestimmung

Our property was a square quarter-section, one-half mile and one-half mile. During our residency in Canada, the country converted to the "metric system" so officially our land was 804.67 meters by 804.67 meters. But this is not quite correct either. The metric system uses the centimeter as the basic length unit, so our land would have been 80,467 centimeters by 80,467 cm. The Systém Internationale (SI), which uses the meter as the standard length unit, is the correct unit system used in Canada and in all the world's countries, except the USA and a few other "third" world countries.

On this land, as I earlier stated, there was approximately an 80-acre parcel on the north, just about a rectangle which was clear and had been used for planting crops. This part was good loamy black ground except for a strip at the far west end that bordered on a muskeg or swamp, and this made the land there swampy and not used for crops. It was no more than three acres. I have sketched the land layout and included it in the appendix. Then, the south half of the farm was a boreal forest mixture: lots of poplar trees, birches, a few groups of spruce, willow bushes, and then lots of mossy ground. There was a nice, big dugout running parallel, east to west to Highway 49 just 40 meters or so from the border with the highway. The dugout, as the natives told us, was maybe 20 years old and very deep in the middle. It was created when the highway project to turn Highway 49 into a passable dirt road needed fill. There were alleged to be lots of fish in the pond, and it seemed believable since it was equivalent to a small lake, 80 meters long by around 25 meters or roughly 0.2 hectares (½ acre). There were cattails and lily pads on the edge, and a number of trees draped their branches over the edges. It was truly a beautiful body of water and was, in spite of the fact that it was dug out by the highway department, a natural beauty. We very quickly decided that our home would be placed overlooking the dugout. And the west end seemed to be a good spot for a home as here the land rose a bit so we could place the building at an elevation about 10 meters above the water level. Not only this, but since the pond was so large and deep, the water would not turn black or become overloaded with algae and other reactions. Most of the neighbors had dugouts that could be good for summer and early autumn, but the water would be dark and distasteful in the winter. Their dugouts were smaller and shallow. So we had everything we needed, and it was all beginning to take shape.

The 80 acres of boreal forest had no trails or means to conveniently get to the location for our new home, when it became a reality. On September 18 we started to clear out for a lane to our future homesite. By the following day we had cleared maybe 40 meters of lane, and it would be tough going. But then, a couple of days later, we had a caller to our rental home. He was from the highway department and told us that Highway 49 was due to become upgraded to a paved, first class highway. They needed lots of good dirt, as he stated it, and with our property right next to the highway, and with documentation that the dirt on the property was "good dirt," would we be willing to let them dig another dugout, somewhat distant from the first dugout. I hesitated and must have given him the impression that I really just wanted things left alone and quiet, with no tree destructions. I didn't say anything like that, and I had the distinct feeling that we right up against the system, Kafkaesque if you will, and they were going to dig it with or without our permission. We were just landed immigrants and rather poor ones as well, but the man kept talking and made a proposal that if they could get their good dirt it would create another good dugout for us and they would clear out a lane for us so we could access our new home from Twelve Mile Road. We thought about that, knowing that we didn't need another dugout, but a cleared lane would be a money saver, not to mention the time needed to clear for a lane. We agreed to let them dig the dirt, not two weeks after we had purchased the land. I wonder often whether George Blakus knew about the plan to rebuild 49 and whether he felt that getting rid of the property before the digging was best. That is all just silly mind thoughts.

Within another week the highway crew was in with their bulldozers, two large four-wheel scoops, and a road grader. They cleared out maybe 1/4 acre of land and dug, going north to south and south to north, proceeding to dig dirt and deliver it to the highway construction sites. They were rebuilding Highway 49 all the way from the Alberta-British Columbia border east to Gordondale, about 20 miles. We were about in the middle, so that too must have made our property attractive for good dirt. Within a week they had a large cavity dug maybe 60 meters running north and south by 20 meters wide. I remember going down to the properties once or twice, to observe the desecration, and the machines were at least 15 meters below the land surface and they were still digging. The cavity eventually was at 30 meters deep at the middle, and while the area was not quite as big as our original dugout, it was big. Over the course of the next two years it would fill up with rainwater and a bit of runoff. The land here was loamy on the top, but about 3 or 4 meters below the surface the dirt was clay, and it was a slick clay too! Water could not permeate the clay very well, and as the water filled the cavity, the pressure of the water on the clay must have made it nearly impenetrable. So we ended up with two good dugouts, if "good" is used quite loosely.

During all this activity, we were trying to decide and plan on what type of building we would need to build for our home and how large it should be. Time was ticking on, and from everything we read and heard, October would be "cool" in the natives' terminology. We were past the third week of September now, and on September 23 we had a hard frost. So while we knew where we wanted to place the home on our property, it was not clear what the home was. Anyway, one day one of the supervisors came to our rental and said they would do our lane, but they needed to know where to do it. We went to the property, and it all seemed natural that just to the north of the new dugout, accessing Twelve Mile Road maybe 120 meters north of the intersection with Highway 49, was the right spot. I walked to the west from the road, visualizing the lane as it would get to the house spot, and fortunately we had already cleared out a portion before being contacted by the highway department. Recalling it all now, I remember thinking what a pity that at least six or seven nice spruce trees and many, many poplars, birch and black poplars would be sacrificed for the lane. A small grove of spruce had already been bulldozed over for the new dugout, and this just seemed so wasteful. But then one can also see how the Europeans coming to the new continent from 1492 on must have thought how so many things were in their way that needed to be removed and destroyed. God had said this was all good, but he was not very practical.... We were being practical, and between a bulldozer and a pass or two with the road grader, we had a path or lane through our woods from Twelve Mile Road to the homesite. There turned out to be a pile of tree trunks piled on both sides of the lane that were standing where the cleared lane now was. This was a quick fence that turned back our horses and cow when that came. The bulldozer even cleared a level spot for the home. As it turned out we would use the help of the bulldozer once more when the home was nestled in its site. So at every turn I had this schizophrenic feeling that I didn't want to destroy a tree, a plant or a home for some wild animals, but we "needed" a dwelling and we needed it as soon as possible. So I was getting lots of help from an eerie development of cooperation and chain of events. I was grateful but with a strong dose of asking for forgiveness.

So the next helping hand came not a week after the highway crews began getting our good dirt. We found out that Willi Meyers had a log home that he wanted moved off his property. Willi was a second or third generation Peace River farmer who had a nice set of buildings just three miles north of our property along Twelve Mile Road. Actually his farmstead was situated about a half mile west of the road up a long side road, and it too was a picturesque setting. I remember that there was a creek running parallel to the side road to the north and which flowed into another large creek about three kilometers to the west through a muskeg, a short bridge at Twelve Mile Road and another timber bridge so one could access Willi's farm on the north side of the creek. This timber bridge was wide enough for the farm equipment used on the farms and was also stout enough to allow a large load to cross. Willi's farm had a modest frame home, built probably 25 years before, with the common tri-color paint scheme. It was a single story dwelling with maybe four bedrooms and the other typical up-to-date arrangements. In the back, to the north and toward the west was a corral and a large wooden classical barn. It was probably 60 by 40 feet and had a cavernous hayloft, defined by the picturesque arched roof, in the vein of a gambrel rood but smoothly curved from the eaves about 15 feet off the ground to the ridge about 60 feet in the air. The barn walls were horizontal boards, overlapped, and they had not been painted, so they had that rustic brown wood look, very nice.

And directly behind the home with a 30 meter space between sat a 32-foot-by-25-foot log home. It was a real full log home with a second floor and three dormers. The house faced south, but the main entrance was through an addition, a 20-foot-by-25-foot wide shed type building on the west side. So the main entrance to the log home was on the west. There were two dormers facing north and one south. The house had a kitchen when one came in and a small living area to the right going to the far side. A small bedroom was nestled in the far left corner of the house, and a stairway started on the far left side making a sharp right 90 degree turn halfway up to the second floor. There were two bedrooms upstairs, separated by a small hallway. The single dormer looked out from the hallway, and the two other dormers were for each of the two bedrooms. The building was much smaller than we had planned on, but things were going in a direction and development that all seemed preordained. The building was actually ideal, perfect, and beautiful as an afterthought. It was a full log home, and the roof was cedar-shingled. As we found out, long before modern engineering and architectural theory has pummeled human logic into "science," the builders had put a foot layer of cedar chips between the roof and the ceiling of the bedrooms and hall. The dormers were framed in, but they too were insulated. There were four single-layer, double-hung windows on the first floor, one over a kitchen cabinet in the center of the kitchen outside wall to the north, one large window on the south about in the middle of the long wall, and then two more at the east side, one in the small bedroom and the other at the living area. Upstairs there was one window at each end of the house, in the middle of the gable, and then three other windows one each in the three dormers. See the appendix for a floor plan of the house.

Finally, what I really became fascinated with was the walls, all made with beautiful large logs, probably first-growth pine. One could see that the logs were taken from large trees because the widths were anywhere from 30 centimeters to 60 centimeters. And all the logs went from one end to the other so they were either 20 feet long or 30 feet. One would be hard-pressed to find one tree today that would furnish a log for such a project. But in addition, the logs were fashioned with a double dovetail at each end so that there would be a tight fit at each corner between the intersecting logs. See the Appendix. Between each of the logs was a caulking of forest moss. Some of the natives called this a "Norwegian" log house, and that may be so, but what difference did it make? It was surely constructed by people who knew how to build a log home. The typical log home today is an overlapping of the log at the corners so they protrude out, like a "Lincoln Log" toy house. This house was clean-lined. When we saw the building, the inside surface of the log walls had been covered with a hideous panel board arrangement, clearly indicating that some of the occupants were embarrassed with the log house and wanted to cover the logs up, even if only on the inside. The house had stood vacant for probably the past 25 years, as Willi Meyers and his family, or maybe his parents, hurriedly moved out of the old log home to a real home. The log home had no running water and no inside toilet, nearly unlivable. But here again, I need to be forgiven. Somehow Willi or his parents didn't burn the old home down, and they were ready to sell it to us. Thank you.

There were a few glitches in the building: With no running water we had to plan something, with no electricity we needed some lights, and I could see some minor structural deterioration. At one corner in the kitchen a bottom log had rotted at the bottom and some of the caulk between the logs was coming out. But all this was just minor laundry, and everything was just taking a self-directed path. We dithered a bit on the value for the building, but $2,500 was the final agreed price. Included in the sale was the shed addition and a nice sturdy two-hole outhouse, both of which would serve us well. I was happy that I now had something the work with, and it was now up to me to get things together....

CHAPTER SEVEN
The House Move

We had accomplished quite a bit during the short time that we were in the Peace River Country. We had met a lot of very nice helpful people, had a quarter section of land, and had a log house to put on the property. We had the lane cleared to the house site, and the house site itself was ready.

The most pressing issue was how to get the structure moved from nearly four miles north of our building site. It was a house-moving project, and as it turned, this too fell into place. The Mennonite Church where we were fellowshipping had acquaintances with another Holdeman Mennonite Church in Silver Valley, and among this group was a member who was a professional house mover, Leonard Johnson. So on a Sunday afternoon we drove up to Silver Valley, about 20 miles north and east of our new property. We had the address for Leonard Johnson, and when we arrived at his home we were welcomed warmly. We had called before, and the Peacheys, members of our church, had also contacted them to let them know that we were coming. We spent a few hours with Leonard and his family, and it was very enjoyable. I remember that they were living in an older home but very comfortable. They let us know that they were planning on building a new home and destroying the old one. Tear it down, so to speak. Well, I was mostly interested in Leonard's opinion of whether I could move the log house and any helpful hints for me. He was a comfortable talker and told me that I could move anything with enough power, so a big tractor would be a good candidate for a 30-by-20 house move. He told me he had moved the house that we were in right then from southeastern Saskatchewan, a distance of probably 2,000 kilometers. He had dollies or sets of wheels to put under a house for long-distance moving, but he recommended just getting a pair of good-sized logs, maybe 2 foot diameter by 45 feet long to drag the house home. He suggested putting a substantial timber at the back of the house, from where it would pulled, nailed to the two skid logs, and then wrapping a heavy cable around the timber after sliding the cable under the house from the towing end. He volunteered to come out when I got ready to move the house and make any further suggestions. That was a very insightful, thoughtful and pleasant thing to do. I indicated that I would be pleased to give him some consulting money, but he declined; he just wanted to be a good neighbor.

Almost like a dovetailed or just in time planning by someone other than myself, Peter Kut, a very nice neighbor from near the border and north of our rental home, volunteered to move the house with his big four-wheel-drive Versatile model 835, 1909 horsepower tractor. Not to be outdone, Marvin Meyers, the brother of Willi and who lived next door, south and across the creek from Willi in a large house trailer, volunteered to use his four-wheel-drive John Deere 7020, 135 horsepower tractor to move the house. So as I found out rather indirectly, Marvin had wagered Peter that his John Deere tractor could move the house by itself. It was a size smaller that Peter's Versatile. So I had two very good neighbors volunteering to move the house. At all chances, I needed to pay them for their efforts, but this was a once-in-a-lifetime opportunity for them to demonstrate their prowess. Not only that, but my budget always seemed to get helped along with generosity.

But I needed now to get the house ready to move, as there was no time to lose. It was late September, and cold, snowy weather was coming. One could feel winter in the air, if not on the ground every morning, and the evenings were getting cold as well. The ice had not yet formed on the dugout, but the children were looking forward to ice skating on it, so I needed to really get to it. The first frost occurred on September 23, 1978, and the leaves on the poplars were turning bright yellow. The birches too were with full dark yellow leaves. My good friend and church preacher, Jesse Peachey, was well-seasoned in the logistics of setting up house. He offered to install a waterline from the dugout to the house and also dig a drain line and soup out a lagoon. The waterline went well. He used a PVC tube with a foot valve on the intake end, dropped in the dugout. The other end was buried 6 feet underground to try to prevent freezing of the waterline. He seemed to think this would be good provided the ground was well covered with brush. As a hedge, we wrapped insulation around the line. Jesse dug a shallow root cellar in the foundation area and also a trench to just outside the house and told me he would come back later to finish the drain trench.

On September 27 Jesse Peachey dug out for the house with his backhoe, and this saved much time. I hand dug a foundation trench, mostly about 2 feet deep around the perimeter of where the house would sit. I formed for a concrete footer, knowing that I was now putting this foundation right on top of the ground, mostly. The neighbors all told me that I would get by nicely as the ground was usually dry and winter ground expansion was minimal. A 2-foot-deep foundation was fine. So I filled the trenches with lots of native rocks to add volume and strength and to economize on the amount of concrete needed. We poured the foundation over October 6 and 7, and in my records it seems we spent $192.89 (Canadian dollars). We used 28 bags of cement, 10 square yards of gravel, and I had to buy a pulley and belt for a borrowed mixer. The waterline passed under the foundation and was exposed inside the house area. During all of my work I was helped by neighbors and friends. Jim Hackman helped with the cement work, and that always can use extra hands.

During our efforts our two draft horses, Bess and Morning Star, were mostly on the back burner. I was intent on working them, but that needed to come later, after we moved to our own home and farm. But they did require some attention each day for feed, water and just a conversation so they would not forget me. And they would need feed for the winter. On October 14 I baled 109 bales of barley straw that I was given to use. It was in a field diagonally across from our land on Highway 49. Again, our good friend Jesse Peachey let me use his baler and a tractor. I had hoped to use our John Deere A, but it needed a new clutch, which we did subsequently replace. I did use the B to rake the straw and some dry hay that I cut with our pull mower along the highway. Here the highway had a very wide shoulder and grass space. So I went ahead and made about five passes with the mower along our highway frontage. Between this hay and the barley straw, we would get through the winter, we hoped.

And I negotiated with our renter to have a small granary, a log building with a decent roof and wooden floor. It would serve nicely as a shop after we became more settled. This gave me four buildings to move: the log house, the shed addition, the outhouse, and this granary. The outhouse was the smallest, and I felt I could move it with my John Deere B. That turned out to be good experience for the later moves even if it went smoothly. Actually, on October 13 I used the four-wheeled wagon, took off the sideboards, and drove the three miles up to Willi Meyer's and tilted the structure onto the wagon bed. It was not a heavy building, only 5 feet by 5 feet and roughly 7 feet high, mostly built of spruce. I had prepared a convenient spot for the outhouse not too far from the homesite, dug a pit, and placed the outhouse over the pit. This was a quick morning project and gave us a permanent structure on the property. I felt like we had made progress. The log granary was moved by Jess Peachey with one his bigger tractors on October 16. My two tractors, the JD B and the JD A, were too small to easily move the building, and since it was almost eight miles away, the trip would have been very long. We put the log granary to the north of the lane but in the area of the home and near where I planned to put up a barn. So in two days we had two buildings on the property.

On October 21 we were ready to skid the lean-to building to our property. It was 10 feet by 20 feet and was now a three-sided building, as it had been placed up against the log house. We had two 24-foot-long logs under the building to skid it. The next day we took our John Deere A, with a new clutch plate, and pulled the building home. We were feeling like old hands at building moving, but the test would be the log house.

Meanwhile, I was working every day to get the log home up on stilts so I could then slide log skids under it. The first task was to separate the shed addition from the log house, and this took a day or so of rough work. It was more good experience though, since I needed to jack the building up after I had it separated from the log home. I had two Jackalls, or heavy bumper jacks. I also had two hydraulic jacks, and Peter Kut loaned me a locomotive jack. This one came in handy for heavy lifting. My set of four jacks was more than enough to get the shed up off the ground, and I then put two log skids under it and used the John Deere A tractor to pull it over to our farm on the afternoon of October 21. We placed it near the homesite, as I planned to use it as our entry, mud room and pantry for the first winter. That turned out to be a good idea.

The weather was holding out right now, cool, or maybe cold, but dry and not windy. During our time in the Peace River Country we did not experience windy weather; it was just mild breezes, if there was air movement at all.

And now I was working against the clock, the calendar and nature. At least I felt that way. I looked at the calendar and made October 28 the goal to move the log house and set it down.

The log house was heavy, and as a way to lighten it up a bit, we took out the original chimney. It was a concrete one, supported at the west end of the house, in the kitchen area about three feet off the floor on a shelf. It took a bit of beating, but we got it all out, and that left a hole in the roof. When we finally had the house on its new foundation, I installed a new insulated stainless steel chimney which worked real well for both the kitchen stove and the heater.

To begin jacking up the house, I needed to collect a large amount of wooden blocks to hold up the base as I jacked it off the ground. As I quickly found out, I needed plenty of blocks under the jacks, too, to keep them from sinking into the ground. This was a real engineering statics project and while the theory was straightforward, the application was tedious, slow and potentially dangerous. Starting at one corner, where I knew the ground was reasonably solid, I dug out a pit deep enough to put the lip of the bumper jack under the corner. Surprisingly enough, the corner actually moved up a bit. Of course, the jack sunk into the ground at least as far as the house moved out, but I could see that everything was able to be done. After lifting the corner up maybe 15 cm, I placed a block under the log and let that corner rest on the block. I proceeded to raise the log halfway down the long side of the house. Again, I dug a shallow hole under the log, inserted the jack lip, raised the log and slid a block under it so the log could rest on the block, instead of on the jack. Working this way around the house, I was able to get the building 15 centimeters or so off the ground. But I still had a long way to go, and it took time to do it.

As we found out, the weather in the Peace River Country was very steady. There would be long periods of dry, sunny weather, then another long period of rain, or snow in the winter. It was our fortune to have a nice warm, sunny and dry period to get the house ready. This was all working to our favor as the ground was not so wet that the jacks sank into it far. If we had had a rainy season, the moving would have been most difficult.

I needed to get two long skids to move the house. I had done some exploring in the Saddle Hills just south of Highway 49 and up the rough trail that took off from the intersection of Highway 49 and Twelve Mile Road, finding piles of spruce logs cut down by oil pipeline installations. On October 23 I took the JD B up the hill and chained one of the largest logs I could find to the back of the tractor. I then dragged it to the log house ready to be placed under the house when I had it up high enough. Repeating this haul one more time, I had the two skids ready for the move. The logs were about 2 feet in diameter and 40 feet long.

There was also a need to get a heavy cable and this I rented, about 25 meters of 2.0 centimeters diameter braided steel cable from Northern Supply in Dawson Creek. This seemed to be normal in the oil drilling industry to rent heavy cable. It saved me a lot of money to not have to buy the cable, and it would never be used by me again anyway. I did need to purchase two clamps to hold the cable after it was wrapped around the skids. I think I may still have one or both of them in my pile of things.

And every chance I had I would make a trip around the log house, jacking it up a few centimeters, inserting another series of blocks of wood under it and go on. I had eight piles of blocks now under the house, one at each of the four corners and midway under each of the four sides. I was piling up the blocks, and the log house was slowly rising up.

Early on in the process I found out that the floor was not usable. It had rotted badly and was not attached to the logs, so I mostly ignored the floor. This was good since there was a small basement under the kitchen portion of the house, and the floor just covered that up. There was a floor trap door that I opened once or twice and noticed that the basement was full of water. That was no surprise since the ground was able to hold lots of water in the dugouts around the area. Basements were just other dugouts ready to fill up with water. I did wonder if the folks who built the house and lived in it used the basement and if it was dry. Just another bit of thought. But with no floor I now needed to get some boards for a floor. I asked around and found out that some of the people had planed some 2-inch-thick tongue-and-groove lumber. This would be just the right material, strong, and give a nice tight floor, but who would have some left in late September? I could always go to the lumber yard and buy some tongue-and-grooved 1-inch boards, but this would require floor joists every 16 inches or so. That could really become an expensive proposition. Finally I heard that Ernie Poratz had some tongue-and-groove lumber. Ernie lived about four miles north and then four miles east of us, not so far. We went to his place and found out that all he had in 2-inch-thick lumber was 6-inch-wide boards, and it was not tongue-and-grooved. It was getting late in the season, and the thought occurred to me that I could install floor beams, rough logs actually, 2 feet apart and then put down half-inch-thick plywood and cover that with Ernie's 2-x-6 lumber. It was smooth and would make a good floor. So I went on and bought enough lumber for the home floor. The second story didn't need any flooring as it was there, but the kitchen, living area and downstairs bedroom did need a floor.

After a 10-day effort, putting me in the third week of October, I had the log house up 65 centimeters off the ground on blocks. It was nearly ready to move. I still recall the night before I installed the log skids, October 26. Mostly the weather cooperated nicely, and it was dry and warm enough for that area. And normally the winds are very mild in the Peace River, but this night, for some reason, there suddenly developed a real storm with rain and high winds. I lay in bed hearing the wind and rain even as I slept. I could see the log house falling off the eight piles of block, nullifying my past weeks' efforts. But again, when morning came I hurried to see the house, and it was quietly situated on top of the eight piles of block, waiting for me to install the skids. That required some extra effort as well, but by the end of that day the log house was resting on the two log skids with the large end to the back and with the heavy braided steel cable ready to be hooked to a tractor.

The next day, the moving day of October 28, Marvin Meyers was there with his John Deere four-wheel-drive, and Peter Kut was standing by with his Versatile. The weather was cooperating nicely, and it was expected to be that way all day. The news of this event had been well disseminated in the community, and there were many people present, maybe 50 or so, all curious to see if the four-wheel tractors could pull the house. I remember a number of cars parked along the side road, and I think an enterprising person could have set up a lunch stand and made some money. Leonard Johnson came to look it all over and seemed to be sure that there would be no problems. The local power company people were there because at the Willi Meyer farm there was one feeder power line crossing the path that the house needed to take before crossing the bridge and on to our farm. They only required that someone get on the roof of the house and with a long stick push the power line up so the house would clear. It was all fairly straightforward, but today some safety experts or whoever would have had kittens seeing the sight of a 20-by-40-foot log home getting ready to move under the power lines, across a wooden timber bridge and then negotiate a sharp left turn before heading east along the side road.

I left the move in the hands of my friends Peter Kut and Marvin Meyers. They decided that Marvin's John Deere would hook onto the house to try to move it off its original foundation. In any case, after a short movement, Peter would hook his big red and orange Versatile to the front, and together they would move the log home to our farm. So, Marvin hooked up with the steel cable clevised to the rear hitch. He was careful to check everything out before he climbed into the tractor cab and slowly applied power. The engine barked, and without hesitation, the house moved ever so slowly and then rather smoothly. It cleared the power line without needing to lift the line, and Peter hooked up his tractor to the front of the John Deere before together they crossed the bridge and negotiated the sharp left turn. This turned out to be a bit of a problem, and fortunately the lane to Marvin Meyers' buildings was right there, so Peter drove up the lane a short way. He then disconnected, and Marvin was able to negotiate the sharp turn with the house behind him. Peter re-hitched after the house was headed to the east. There was another tricky spot at the intersection with Twelve Mile Road, where a sharp right turn needed to be negotiated. But this one turned out to be less sharp, and the two tractors skidded the house smoothly around the turn.

The road surfaces were all mostly dirt, not gravel, so the wear would be minimal. I estimated that the house weighed 30 tons and the drawbar pull at the house was 14 tons. So the friction was 46.7 percent of the house weight. The tractors were able to move at about 1.5 miles per hour on straight stretches, and it finally took over three hours to travel the four miles from the old house site to the new one. I noticed as the tractors turned into our lane and the house following, that one of the log skids had worn down to less than a third of its original diameter. But the house was there. I had not planned well enough to let the tractors move the house onto its new foundations as there were trees on the other side. This was another one of my many blunders of trying to conserve trees. But we had accomplished a significant task: moving the log house to our property.

On October 31 the highway department brought one of their bulldozers to our house and proceeded to winch the log home onto the foundation. The bulldozer operator, Ralph Suek, was able to squeeze his machine between the trees, and with the house up on some poplar poles, it was a smooth process. After all this jacking up, shackling and moving, the house was as square as when I first saw it, the doors were all in line, and no windows were even shaken. It was all testimony to a solid building and good careful driving by my friends. On November 3 we fitted spruce logs for floor joists on the first floor. We were now inside the building, but we had no heat and no light, so we were limited to just a few hours during the midday for work. But by the end of the day we had the house down on the foundation with floor joists installed, ready to install flooring.

Then, the next day we received the first significant snowfall. It snowed throughout the day, but by evening there was only 2 or 3 centimeters on the ground. But we now knew it was coming into winter, and we still had a lot to do before moving-in day.

During the next three weeks we were very busy. We had suffered through over two months of living in the Tomzeck place, and with the cold weather, all the inadequacies of that bare arrangement was getting even worse; the small water pond was icing over now, so water was hard to get; the kitchen cook stove was the only reliable heat to the small house; and the cramped conditions were wearing us all down. We needed to move into our new home. And this was a true motivator. Almost every day we drove down to the property and worked. Kurt put off going to school so he could help with all the various details that needed to be completed. On November 15 Kurt and I drove from the Tomzeck place to our property. It was a warm and sunny day, just perfect for working outside. We drove on Highway 49 from our rental to the intersection with Twelve Mile Road, where we turned left. With the dry weather, the roads were dusty, and fast vehicles made more dust. Kurt was driving, and I was sitting in the rider's seat of the one-ton truck. He turned into our drive, just 120 meters or so down Twelve Mile Road, and didn't quite make it to the lane, because I heard and felt a hard impact on the driver's rear side. The truck, ever so slowly, tipped over on its right side, and Kurt, with the help of gravity, bumped against me. I was sandwiched between Kurt and my rider-side door. There was a lot of dust, but we were both OK and there was no flying glass or other obvious damage. To exit the truck we had to climb out of the driver's side door. Kurt went first and then I. It was almost surreal to get out of a truck by climbing up and then out. But when we were out we could see a pickup truck enveloped in dust and sitting just a few meters away from our truck, stopped. It was clear that the pickup truck had hit our truck as we were negotiating the turn into our lane. And while our truck was lying on its right side, it was not badly damaged. The rearview mirror on the rider's side was the only real damage. That was not so of the pickup truck as its front grill and hood and probably frame were crumpled in. The pickup had hit our truck right at the most rigid part, on the steel bed frame. So the give was all in the pickup, and that made it mostly a total loss. Not only that, but the windshield was shattered, and three people were still sitting in the cab, side by side. The driver was a middle-aged neighbor from a few miles north and east, in the middle was young boy, and on the right side was the driver's wife.

The wife and boy were both crying because two rifles had fallen from a rack at the back of the cab and hit them on their heads. The driver, when he started, was mostly concerned about the integrity of the two rifles, forget the wife and son. And he quickly got out and surveyed the damages. His truck was not going anywhere soon.

I thought the first thing we needed to do was contact the RCMP, Royal Canadian Mounted Police, so I went to the Pilsworths' across Highway 49 and used their telephone to call our accident in. We were 70 kilometers or so from Spirit River to the east and 50 kilometers or so from Dawson Creek to the west. But I called into Spirit River because that was our jurisdiction. About half an hour later the RCMP came, probably driving very fast to make the trip in that short time. The officer was very courteous and did all the normal procedures: drivers' licenses, accident experience and survey of the vehicles. We were surely not in the wrong as Kurt had a new driving permit and the other truck was in the wrong lane when it had hit us. We even had vehicle insurance with the Coop out of Dawson Creek. But we were "foreigners" to the other driver. So he must've used that plus the normal behavior of trying to say the other person was at fault. He pulled out the statement, "They didn't have their turn signal on when I was going around them." The officer diplomatically asked Kurt if he did have his turn signal on. "Yes, I did have it on. I was making sure to obey all the requirements of safe driving." We all walked to our truck and noticed that the turn signal was still blinking, even after the impact and tipping over. With that, the officer quietly wrote up the accident description: The pickup truck was traveling at a high rate of speed on Highway 49, rounded the corner onto Twelve Mile Road, and attempted to pass on the left our truck which was making the final turn into our lane. Our turn signal was blinking, and we were moving away from the pickup truck. It was all clear and standard.

The driver had a "mobile," or what is now called a cell phone, and he contacted some people who arrived shortly after the RCMP and proceeded to drag their disabled truck home. We used one of our tractors to pull the truck right again. It had a high stock rack on it, which made it easier for the pickup to tip our truck over, but nothing was seriously damaged, and we finally were only out a few hours of time and a bit of an experience that we would rather not have had. As it turned out, a few months later, in March, Kurt and I had to appear in a courtroom in Spirit River to testify about the accident. It seemed that the insurance company of the other party, which was the same company as ours, wanted to cover all its bases. So Kurt and I had a bit of a courtroom experience as well, again losing some precious time.

So we continued with our work in getting the house ready for living in. I remember patching the bottom corner of the house by the kitchen area. A log had rotted away, so I replaced a section. In addition, some of the chinking, moss "caulk" if you will, needed to be replaced, so I went into the woods by the house and scooped handfuls up and took it to the house. I found the caulking of cracks between the logs was an easy, natural and satisfying task. The house roof looked good, and all the windows and doors were intact and square even with the dragging of the building some four kilometers down a rough road. The house was solid and square, and by the morning of November 24 we were ready to move. We had a drain line dug from the kitchen sink to a hastily dug lagoon some 50 meters away from the house in the trees. It was getting cold now, and I remember that we had to build a fire along the surface of the ground where we planned to dig the drain line. Jess Peachey, with a small backhoe, did the work of digging the lagoon and the drain line from the house to the lagoon. Before doing the drain line, we connected the waterline from the dugout to the house, where we had a little green pitcher pump, purchased at Quality Farm and Fleet in Greenville, Ohio, in the kitchen cabinet overlooking the dugout. The floor was supported by log joists running across the house, and these were covered by plywood, and then 2–x-6 planks were installed. It was -10 degrees F on the morning of the move, but who was worried? There wasn't much snow yet, and we had a small pile of wood at the log home. With everyone pitching in, we negotiated the move to the log house by the evening of November 24. We awoke on November 25 to at least a foot of snow on the ground and -40 degrees F temperatures. But we were in our new home, and it was comfortable compared to what we had experienced before.

Joy was to write about our new home and settling in on December 3, 1978.

It has been 8 days now since we moved into our log house. We put down a new floor, 2 inch x 6 inch. The tongue and groove flooring that we wanted was not done and we could not wait for it. We varnished the boards twice, but after the move and tracking in snow and soot from the stoves, the floor needs a good scrubbing and some more varnish.

The house is easy to keep warm, it is well insulated with the outside chinking of moss, which is the best insulation available. We have two stoves downstairs and we opened up the stairway which had paneling to close it off. Now much of the heat goes upstairs but there is enough for both downstairs and upstairs.

A pitcher pump supplies us with running cold water which is heated on the cook stove. We bought a second hand stainless steel sink for $18 which is very nice luxury considering that there is even a drain *which runs to the lagoon...*

...It is a bit crowded since the lean-to is not against the house. I planned to do the washing in the lean to but now I will have to do it in the kitchen and hang the wash outside. I have cut down my washing to once a week , making it a little less effort. There are a lot of clothes but not like in the summer time. I plan to wash out the diapers by hand twice a week and hang them out.

We are very pleased with the house so far, it is quite comfortable, but we do have some extra uninvited guests living with us, namely pack rats. They are quite persistent little rascals and insist on making their living quarters in the kitchen cupboard. We have to destroy their mattress-filling nests almost every day. There are some old mattresses laying outside which were left in the log house and they must be pulling the stuffing out of them. All the food has to be secure also, because they play smorgasbord otherwise...

Our sixth child is due any day now and we are all anxious for him or her to arrive, especially the children. We don't have insurance so this will come out of our pockets.

We have had lots of visitors lately since we moved. Jim Sadlier, Paul Sadlier, Phil Sadlier, Jesse Peachey, Willard and Arla Stutzman, Nick and Rose Kanda, George and Arlene Blakus, and Fred And Irene Pilsworth.

Tonight everyone went to church except me. It is getting harder for me to ride and sit for any length of time, so I mostly just stay home and keep the fires going.

I have to make my curtains yet, as soon as we find the curtain rods. I have some pretty material for the kitchen and can't wait to get them made and hung. Kurti fixed up my old Davis treadle sewing machine and it runs very well. It makes very neat stitches and a does a fine job of sewing. I am happy that Kurti is such an obliging and handy son with fixing things for me. He is a rare and talented boy. Lori too is a big help to me. How we rated such clever and ambitious children I'll never understand.

The children are going to sing at a local hospital this Wednesday night. They will sing songs that they are planning to sing at the school Christmas program on December 20th. It sounds very interesting although we are not supposed to know exactly what they will do.

Things are shaping up, thank God...

Joy would have her baby the next day.

Finally, our oldest daughter wrote a poem during our settling into the new home.

Our Home
By Lori

I think our home
is cozy and nice
it's pleasant for everyone
even the mice

It's happy and lively
with kids all about
they run and laugh
and yell and shout.

A happy home
means to me
a home like ours
A pleasant place to be.

When I read this poem it surely brought back fond memories. And curiously, Lori wanted to make it have some "seasoning," so she tried to just burn the edge of the paper a bit to make it even rustic. Somehow, she burnt it enough that the last two words of the poem (to be) are missing, maybe?

CHAPTER EIGHT
Our Sixth Child is Born

Our children were our strength. At all turns they helped in so many ways it cannot be described. Much of the help was just moral support and encouragement, but the older ones in particular worked hard and uncomplaining with all the tasks they saw that needed to be done: washing dishes, washing clothes, cleaning floors, putting wood in the stoves, cooking meals, and on and on. The two oldest, Kurt and Lori, were just steady with helping and also being good sounding boards. We were on an ambitious adventure or experiment, and they helped guide the way. Doing carpentry work, setting up the clothesline, and all the many obstacles we encountered were smoothed away by them.

And our routine quickly became one where the oldest four children went to the Baytree Mennonite School during the week, and we went to Wednesday evening singing and Sunday church. Sunday was always a day of rest, although I found myself often cutting a pile of wood to tide us over. So they had a somewhat normal routine, making new friends, interacting with their teachers, and observing Sunday. At home they had to help out as well, but it was not mind-numbing in any way, I think. At that time in Canada, and even now, the neighbors often came over for a visit and had coffee, tea and treats. This occurred even when we lived in the rough Tomzeck place, and I think it made a new dimension for the children socially. We early on had no television or even a radio or records to play. We did have the piano, and that provided a nice outlet for singing at home, with the children and with guests too.

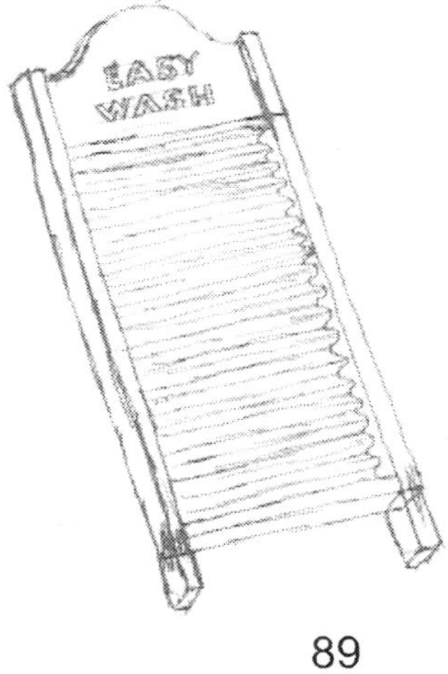

The two middle children, Tim and Heidi, were uncomplaining and helped out when they were asked. Tim tended to be the intellectual type, who was reading anything he could find, was a very good Bible scholar, and provided a steadying influence on us all. Heidi could have been named Charity or Love because we had a daughter in Heidi who always wanted to please others. She also had a gift for art, painting, sketching, writing and even music. Later, when we moved away, she would take up the French horn and become very good at playing it.

Our youngest child, Charity, was apropos. Like Heidi she wanted to please and often silently helped out. But Charity was born just shortly before the organized churches decided that "charity" sounded too much like a welfare program and they surely didn't want to have welfare programs in the organized churches. I believe that the change from "charity" to "love" in the various references in the Christian church was more sinister that theological. But Charity was one who wanted to help others, and I find that wonderful. She was four and a half years old in December 1978 and just absorbing everything.

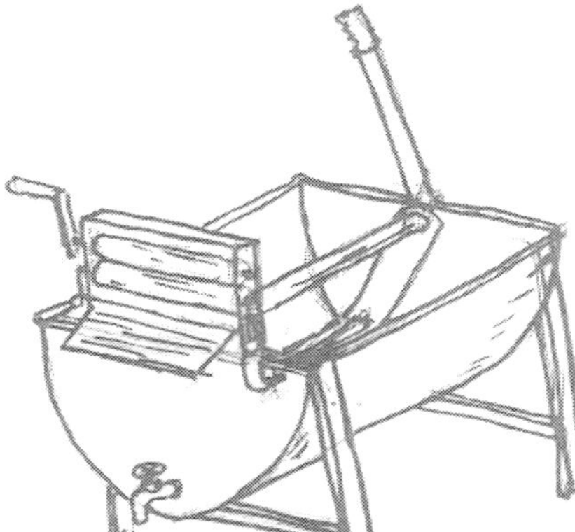

And when we left Ohio in late August, Joy was well along with her sixth child. She was doing exceptionally well, considering all the responsibility and activity she had. It seemed that the whole venture kept her healthy, and the child inside her seemed to be growing healthily. She contacted a doctor in Dawson Creek, Dr. Boronowski, whom she seemed to relate well to. It was pretty much known that early December was her due date, and this too made our move to the new home so important.

Our oldest son, Kurt, wanted his driver's license as he was now 17 years old and qualified for a Provincial driver's license. He already had an Ohio driver's license and helped drive up from Ohio. Having an Alberta license was going to help out. So we decided that Monday, December 4, would be a good day to drive to Spirit River and get Kurt's Alberta driver's license. Joy was ready to have our sixth child any day now, but she assured us that we should go ahead and that she probably was one or two days off from needing to get to the hospital.

Just a few days before this, probably on November 29 or 30, we had really just moved into the log house, and we were always in a hurry to get wood for the stoves. We had a pile of short thin logs that would be good for the cook stove if split in half. Tim and I managed a good arrangement where I would take a short log, place it on end on a chopping block and Tim had a short-handled ax that he used to split the log. This was a case where two hands make short work and even more than the expected halving of the work. It was going just fine when for some reason, probably just a careless maneuver on my part, I got my left hand on top of a log and the ax came down on my index finger. It cut to the bone and Joy had to take me to the emergency ward at the Dawson Creek hospital. They stitched it up and bandaged it, and I was laid up for three days or so. Tim got enough wood for the stoves, I remember, and I was back to normal, with a permanently bent left index finger. Ironically, I beat Joy to the hospital as we knew she was due to give birth soon.

Kurt and I drove the 72 kilometers or so to Spirit River on Highway 49. We took the Chevy pickup truck and left the one-ton. The construction crew had wrapped up the past summer and fall project with a reasonable level grade, but it was still gravel. Past Gordondale the highway was paved but not all that much better than gravel. In the winter, mud was not a problem at all, and whether it was gravel, mud or pavement, it was usually snow-covered and passable. We left in the morning, around 8:30 and by 11 we arrived at the license bureau in Spirit River. Kurt proceeded to complete the written examination and then the driving test. By noon we were on the way home, Kurt driving since he had successfully passed the tests.

But when we arrived home, no one was there. We went across the road, short walking distance, to Fred and Irene Pilsworths' home. They filled us in. Joy had labor pains starting shortly after we left to Spirit River, and she couldn't contact us. So she drove herself to the hospital in Dawson Creek with the one-ton, taking little Charity along. She deposited Charity at Peacheys' and continued. So she must be in the hospital right then, at 3:30 p.m. Kurt and I drove to the Baytree School, which was right on the way to the hospital, and I left Kurt there with the four other children at Peacheys'. I hurried to Dawson Creek and arrived at the hospital at around 5 p.m. I quickly found out that Joy was in the delivery room, and I was given a mask, washed up, and entered the room. Within the hour, at exactly 6 p.m., Sunshine Jennifer was born. With all my stumbling around I was the first one to see her come into the world. Well, maybe Dr. Boronowski gets some credit too. But the birthing was natural and went very well. Easy for me to say; Joy was hurting badly, torn, worn out, and bleeding. But she and the baby were still healthy and robust. I think this was the high point of all of our experiences in Canada, and it did bring us all closer together. We were now a family of eight. I remember the first child we had and I thought then, how is this possible? Why am I so blessed? Who is it that guides me as I blunder along? But I will always love Joy, and this is one of the reasons for that feeling.

After knowing that Joy and Sunshine were comfortable and resting, I returned to Peacheys' and took my five children home. They wanted to see the baby, and that did occur the next day when Joy and Sunshine came home. The house was made nice and cozy warm for Joy and Sunshine, and a good meal was prepared by the children for the welcoming. Outside the temperature was 15 degrees F, and it was cloudy. Snow was holding off though, so that by 5 p.m. we were all together in our new home.

Joy had named our sixth child Sunshine because there surely was not much sunshine in northern Alberta in December or any of the months of winter. But our sixth child did bring sunshine, a light and newness, to our family. It was so nice. But there was also another reason for the name Sunshine. We had a good friend I worked with when I was at the University of Dayton, Greg Williams. He was an engineer, as I was, and he worked in research. During his stay at the University of Dayton he became disenchanted with the research work, which was totally funded by the War Department and was associated in some way with the various national and international conflicts going on. He resigned his position and moved to southern Kentucky with his wife. They both had solid backgrounds in computers, so they intended to do subsistence farming and supplement it with consulting from computer work. And they had some neighbors who were trying to live as "the Amish" but with a born-again theology, or cult. This other family, Sandy and Jerry Tucker, they called themselves "Old Order Mormons" for some reason, had 20 or more foster children and continued to accept more children. We visited with the Williams once or twice, although it was a long journey down from Dayton, Ohio. And we also then met with the Tuckers and found them interesting and friendly. They had just taken in a young girl who was a Native American, an Indian, and this child was *Sunshine*. She was surely a beautiful little girl, and her name did fit. Joy remembered that and said her next child would be Sunshine.

Our five other children hovered over Sunshine whenever they could. Kurt was very good with her, holding her often when she fussed. And Lori was always there to change a diaper or hold her. Tim was ready to help out as was Heidi. Even little Charity was always ready to hold her younger sister. So Sunshine had a good group of siblings and was not neglected. I think back on that and believe that the sixth child was a blessing for all of us.

And December became a good month for us, even with the cold weather. On December 20 the Baytree Mennonite School had their Christmas program. It was held in the evening in the schoolhouse, and our four oldest children were in the program. It was the traditional play with the three wise men, one of whom was Tim, Mary and Joseph coming to Bethlehem, birthing in the stables and the Christ child in the manger. Christmas carols were sung, and the building was full of happy children, parents and siblings. A good carry-in supper was enjoyed and with the wood stove keeping the building nice and warm. It was a real morale builder for us. Our four oldest children were going to the school and seemed to be fitting in well. Kurt was doing correspondence school since he was in high school, but the other three were working with Carol Peachey to learn the basic subjects of middle school and grade school. So the Christmas program was a good demonstration of the preparations for putting on a play and some of the songs they sang during the past four months, and we all just felt good.

Not a week later, on December 29, we celebrated Heidi's ninth birthday at home. She invited three of her friends from school, and it was another good scene for us.

We ended the month with three very cold days, reaching down to -45 Celsius on December 30 and further preparing us for a cold January....

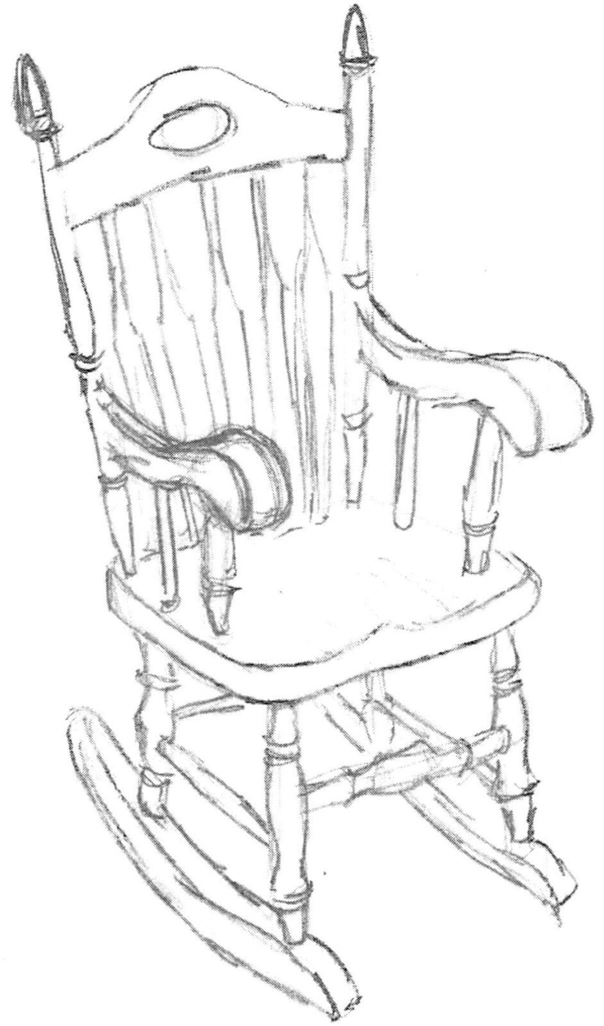

CHAPTER NINE
The Farm

Our quarter section of land was not the typical farm for any place in Canada or the US. Sixty-four Hectares or 160 acres does not constitute a large enough farmstead to support a family for an expected "modern" lifestyle. It would be enough for a subsistence lifestyle, and that is what we had planned. I had expected to eventually farm with the horses, raise enough for us and have a bit left over to sell or barter with neighbors. That is precisely what one should mean by the term subsistence. So we had roughly 80 acres of crop land and an acre or so for the house, a garden, and barn and outbuildings. The rest was woods or bush as the neighbors would refer to it. Some would have called it "trash land," but I considered it the soul land, where one went to truly live and be rejuvenated.

The cleared land had been farmed the year before, having a barley crop, and it was ready for spring planting. I planned on seeding oats as a cover and a cash crop with underseeding to hay. The neighbors told me that cover would be there even if I did nothing and alfalfa was not a good crop this far north. Timothy or brome grass would be acceptable hay, so I planned on following the locals' recommendations. As it turned out, Alsike Clover and Red Clover seemed to grow all along the roadways and any place where the land was not otherwise cultivated. Alsike Clover is a tall plant having small white blooms and was good hay for cattle. Both Alsike and red clover were brought in by Europeans, so they were not native plants, but they sure thrived. Alsike was a bit rich for horses, but mixed with grasses was OK. Red Clover was a more low-growing plant and had large red blooms, very prolific. Dandelions, Burdock, Plantain and quackgrass did not seem to get an opportunity to grow well in the Peace River Country; maybe they didn't like the cold weather. And at the far west end of the cleared land was that strip of muskeg or swampy ground that I sometimes walked through looking for native plants or just enjoying it all. At the homesite, the outhouse was not far off, just to the west on a short path. It was far enough away and yet convenient.

A bit farther away to the west I built a small chicken coop. I mostly put it up in the first winter, whenever the temperatures moderated. That didn't happen until late February and early March. But I was intrigued by the sight of all those tall straight poplar trees, and we had large piles of their trunks along the side of our lane. I decided to build a log chicken house, and I used mostly the chainsaw to make the notches at the corners. The building was small, 12 feet x 8 feet, so I needed 14-foot logs and 10-foot logs. I was able to find enough logs in the various piles, and I worked with Bess, our Belgian mare, to drag them up to the site. I built it as a shed, with one long wall 4 feet tall, the other long wall 6 feet tall, and the two end short walls were trapezoidal. I put a window in the tall wall and a door on one of the short walls. I decided to see how an insulated roof would work out, so I used a pile of saw slabs for the ceiling. Slabs are the rough uneven scraps left over when one saws a log for lumber. For the ceiling they worked great. I put 1-foot logs on top every so often for spacers, filled the space between with wood shavings from the nearby sawmill, and put another slab wood roof on. I tried to make it so there were no cracks to let in rain or snow, and it all worked out good. When we got chickens later on in the early spring, they did very well in their new home and the next winter were quite comfortable. I found out, and should have known it, that the number of chickens one puts in a house is the most important parameter. The right number of chickens will keep themselves warm in the coldest weather, but if there are not enough they get cold and could freeze. How many chickens is based on the chicken coop size and its quality of insulation. It finally comes down to experience and trial and error. But a chicken can be expected to give off about 15 Watts of power as heat, so one could match the heat from each chicken times the number of chickens and set that equal to the heat losses from the building. Give or take a bit here and there, this is exactly how all buildings are sized for heating and cooling systems.

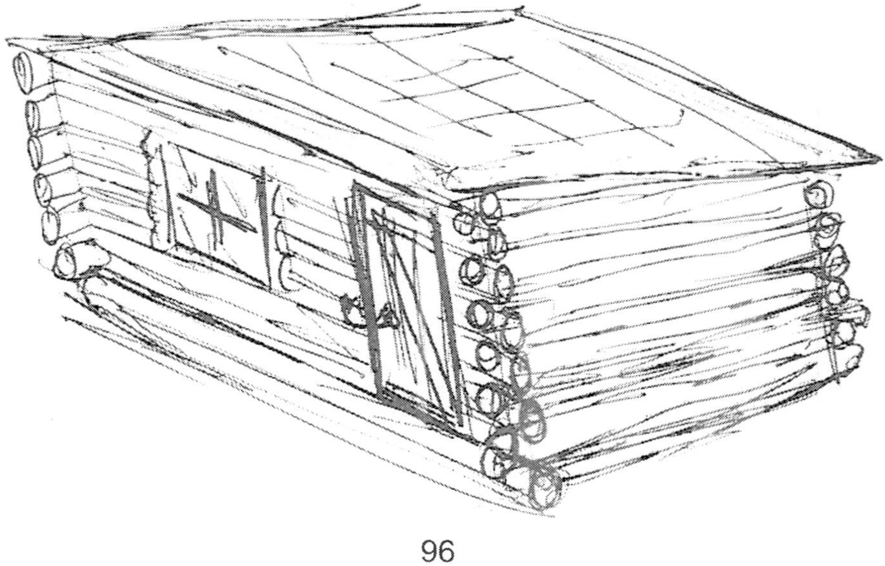

The log granary we got at the old Tomzeck place was set down on the north side of the lane near the home. We would use it as a tool shop, but for now we intended to have a cow, and I could also stable the horses there if it got too cold outside. It was a bit rough, but for now it worked. We also used the A to drag the lean-to back up against the house, where it was at the Meyers place. This would serve to buffer us from some cold, be a mud room, and also a storage area.

Among all the stuff we hauled up from Ohio was a box bed wooden wagon. Joy had bought the running gear near Bradford, Ohio, and I built a box for it, complete with a spring seat and rear-end door to remove grain. I had painted the running gear red, but the box was still bare wood. We had it piled full of miscellaneous items that we didn't need or couldn't find room for in the house. We covered it up with a tarp and mostly left it for the winter.

The inventory of farm equipment we had was mostly old and had been intended for horse farming. That was what I had planned. We had among the equipment a hay mower and grain seeder or grain drill, both of which were originally horse-drawn, and I had the tongues to hitch my horses when the time came. The hay mower was a nice 5-foot cut John Deere ground-driven machine. This meant that the steel wheels drove the sickle bar, the oscillating cutter bar having triangular mower "sections" to efficiently cut hay. I had purchased it in Ohio, following a lead from my father-in-law, and it was a faithful machine in Ohio and would be in Canada as well.

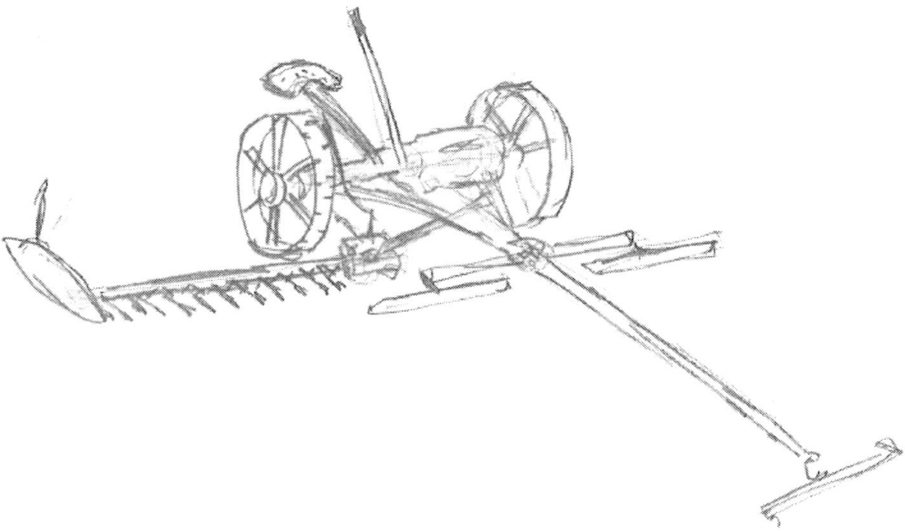

The grain drill was a nice clean, well-maintained machine that would evenly scatter seed, oats, barley, hay seed, or other small grains on a field. It had two large wheels that provided the motion of the various plates rotating inside the drill to drop the seed into the ground. There was a large box on top for grain to be loaded. This machine could seed a 6-foot-wide path, laughable by today's modern farmers, but it worked admirably. Actually I had often seeded hay or rye by hand in Ohio. An acre field could be done by hand. But this meant actually walking the land, carrying a bag of grain and broadcasting or slinging the small seeds on the ground. If one did it with a mild breeze blowing, it was an exhilarating work and healthy too. But the grain drill replaced walking and carrying a seed bag. I would use it with the horses to seed our oats field and hay.

We were short on roof areas and storage space to cover all our equipment and what nots, and it was clear that much of our equipment, the two tractors, the green wagon, the hay loader, the mower, and the grain drill would just have to take the rain, snow and cold. We tried to cover them up somehow, but that was all makeshift. It would also have been nice to have had a wood shed, as dry wood is important. But that too was not a high priority item, so the wood shed was built during the spring of 1979.

During our house moving, we had the opportunity to bale up some barley straw which would be good for bedding and also marginal feed for our horses. It was from a field just to the east of Pilsworths' about diagonally across Highway 49 from our farm. We borrowed a rake and baler and ended up with two wagonloads of straw. We piled this in the barn yard, so to speak, and it helped during the winter. It even was good for the following spring but by then was mostly gone.

As a bonus, in Canada at this time the highway had an exceptionally wide shoulder and drainage field. I think the roadway width was about 100 meters, up to the land for farming or forests. So our parcel, abutting Highway 49, had a 40-meter-wide tract that I considered land that I could farm. And here that land was not mowed or otherwise maintained by the Province. It was not flat but was also not graded with a steep ditch. So I found the opportunity to mow this strip of land, for about a half mile from Pilsworth Corner west. It was rather good hay too, mostly Alsike Clover but also some Timothy and red clover. Here was a bit of oats too, maybe wild oats. But we did end up with about 60 bales of hay that helped over this first winter. I remember baling very late in the evening and feeling good about the endeavor. We would continue this practice the following summer, and we had another 60 or 70 bales at that time. Actually this idea of farming the roadway shoulders was something I had noticed when we traveled through North Dakota and Saskatchewan. So I was just borrowing from more innovate people.

Finally, our horses were needing some shelter, and as I already have written, the log granary that we moved from the Tomzak place would serve during very cold periods to give them shelter and warmth. But they enjoyed the outside, and my makeshift pasture, nothing more than a part of the woods with some barbed wire to discourage their wanderings, served the purpose, along with a long pile of poplar trees that had been pushed over by the highway department when they came for our "good dirt." But the horses were always there, faithfully ready for hay, some grain, or just some attention. They seemed to like our petting of them, and I think it all helped later on when I softened, or mellowed, Morning Star into accepting the work harness and being hitched to the wagon and other implements as well. Horses are surely smart and careful animals who know who their friends are and who care for them, unlike people. So every morning one of my exercises was to check on the horses and give them water and some food. Attention was just a natural part of it all.

CHAPTER TEN
Cutting Firewood

The houses in upper Alberta and British Columbia were often heated with wood. There was an abundance of firewood to be had, mostly poplar or aspen and some pine, spruce, and birch. We heated our house entirely with wood, using an old potbelly stove that we shipped up from the states in the boxcar. It was a good-sized stove, but it devoured wood of all types and could not keep a fire for long times. We attempted to bank the stove at night, that is fill the firebox with enough wood to last a few hours. But as it turned out the stove had sufficient looseness, crackage, if you will, around the doors and at the stove pipe that air could enter easily and literally consume the wood. If the fire lasted for two hours it was doing good. The stove was connected with regular stove pipe to an insulated stainless steel chimney on the east side of the house. The stove pipe was probably 5 meters (16 feet) long, and this pipe did much of the heat transfer to the house. It was mostly horizontal, so periodically, like every other month, it was cleaned out to prevent soot build up. The system worked good.

In addition we had an old traditional cook stove, white enameled body and doors, with a firebox maybe 15 centimeters wide, 30 centimeters deep and 45 centimeters long. It had a common-sized firebox and served us well for cooking as well as providing supplemental heat to help out the pot belly stove. It was placed on the east side of the house and so had a short stove pipe to the same insulated chimney that serviced the pot belly stove. The cook stove had a cooking oven adjacent to the firebox to bake all sorts of breads, cookies, meats and other food. It also had a warming oven above the stove surface for keeping goodies warm until time for a meal. On the right side of the stove, facing the stove front, was a square water tank to warm water. This was an important part of the operations as it had plenty of warm, even hot, water when the cook stove was being used. So we needed plenty of wood. As those who may know about amounts of wood, we needed around 20 cords of wood to adequately heat the home and also for cooking. A cord of wood is literally a stack of compact wood 350 centimeters by 350 centimeters by 700 centimeters, and that is only volume. A cord of good quality firewood provides a lot of warm evenings, but a cord of wet, green wood is not worth much. Remember that the log home was

20 feet wide by 30 feet long, not large, but it was two storied and had nice loose, rattley single-pane windows. This made it difficult to keep the first floor of the home warm. The second story, two bedrooms each with pretty dormers looking out on the "lake" and a hall, was covered with a 15-centimeters-thick wood-chip-filled roof. The roof was sawn cedar shingles, and the second floor was cozy warm most of the time. The five oldest children slept up there and often opened one of the two windows in each bedroom to cool it down, even when we had -50 C or so. This did not please me since I was mostly the one who needed to cut the firewood, stack it as best I had time for, and put it in the stoves. But looking back, it was just part of the home and makes me smile about how I would have been just like my children.

As it turned out, I also had a smaller pot belly stove in the tool shop which I occasionally fired up during the days when I was outside continuously for up to six hours at a time. Also, during the second year after we attached a small building to the entrance to the log home, we installed another small stove in that room. It was fired up very occasionally.

So I needed to get firewood for the stoves. One of our neighbors, Nick Kanda, who was a native to Baytree and northwest Alberta, suggested that the best firewood was year old poplar, cut green into blocks in the fall, 40 to 60 centimeters long and split in half lengthwise. Left to season for one winter and a summer, this wood provided a good hot fire and lasted reasonably well. But I had not the luxury to wait a year for firewood. I needed it now, and plenty of it. I found that spruce was good to start a nice hot fire, but unless you stayed right by the fire, it burnt just a bit longer than paper. Pine was somewhat better, giving good heat and lasting a while. Even the pine needed to be replenished every half hour or so. Tamerack, or western larch, was probably the best as it was a "tannenbaum" but lost its leaves or needles in the fall. It burnt reliably, starting and sustaining a hot fire, and the fires could be exceptionally hot. We had a fire one night in the pot belly with an armload of tamerack logs, and the sides of the stove got red. Tamerack was mostly found in the muskeg, peat bogs and swamps, and was actually not so plentiful. But black poplar or cottonwood was also available, and this we burnt a lot of. Willow was there, but it was mostly small wood and one was never sure that the willow was dead, so I didn't seek out willow for firewood. Actually, I always tried to get only dead wood, as it had a good chance of being dry. And in northern Canada the people mostly considered trees and wood as a nuisance unless they were burning it. The spruce and pine were of commercial value as lumber, but that is another story. So there were many piles of dead trees stacked along the roadways and pipeline trails in the woods or muskeg, which was the common term for the woods. As it turned out, my biggest problem was just finding the time to devote to cutting wood.

One technique that I used was to the cut logs into 3 meter lengths in the places where they lay. Those about 6 centimeters to 12 centimeters diameter were the best; they were not too heavy to drag, lift and load, and they could mostly be easily split into chunks for either of our two stoves.

I would load them on the small trailer hitched to the JD B tractor or in the wagon with the horses hitched to it. I would then haul the trailer or wagon load of fire logs up to the house, pile them up and, when I had another block of time, or then if it was working out, buzz the logs with the buzzsaw. This was a circular toothed saw, about 60 centimeters in diameter on a spindle and frame attached to the JD B. A flat continuous belt was attached to the power take-off pulley of the B to another pulley at one end of the saw spindle. This was an impressive operation and cut firewood in short order. But it was fraught with danger, and one always needed to remind themselves and others to stay away from the saw, as cutting an arm, a leg or any part of one's body could be effortlessly done by the tractor. So I always was relaxed when the buzzsawing sessions came to a close.

This left me with a large pile of logs and some of these needed to be split. The pot belly stove liked large chunks, but the cook stove was best controlled with uniform-sized 30 centimeters long by 6 centimeters diameter chunks--cook stove wood. This meant that some of the logs needed to be split lengthwise, and this too took time and effort. But it all went and was a rather relaxing work mode.

The neighbors who burnt firewood for heat had various stoves. Our nearest neighbor to the north, George and Arlene Blakus, had an air-tight stove. It stood about 4 feet high, 130 cm, and was oval-shaped with a width of maybe 65 centimeters, and it was fabricated of a thin steel liner, probably a 16-gage metal, with enough reinforcement at top and bottom to retain its shape. The firebox had four feet attached at the bottom to put the bottom of the firebox about 15 centimeters or so above the floor. The stove had a damper at the smoke pipe and a damper adjustment at the bottom of the firebox, but the stove was really a barrel with no grates to precisely control air coming to the firebox. It did heat the house effectively and was quick to heat up. With plenty of dry wood in the firebox, it could also hold a fire overnight, maybe seven hours. It always seemed to me to be a short-term solution to heating, but it was a house fire waiting to happen. One needed to frequently check that the liner had not burnt through or corroded. Also the method of controlling the fire, with a damper at the exhaust, made a system which could be unstable; shutting off exhaust gases to control a fire meant that there was the possibility of the smoke backing up and entering the house through the slider adjustment. There were other neighbors who had this same type of stove.

Some people had a larger good stove in a basement with a fan to force hot air, heated from the wood fire, through ducts and distribute the heat to the upper floors of the house. It was a variation of the typical oil or gas furnaces but using wood. The Peacheys had a stove like this in their basement, and it heated their house adequately. As a "benefit" the wood was stacked in the basement through a chute from the outside, and this eliminated getting the house "dirty."

But most of the wood cutting and processing was a just-in-time operation. I remember that we really had no good blocks of time for getting a "winter supply" of wood. One of the earliest sessions involved driving down the main highway, at the time only a mud road, in the one-ton truck with my chainsaw. As I said earlier, there were piles of trees and logs along the roads, and this was considered trash. So if anyone wanted it, it was theirs. There was an early snow, probably in October, and we were still not in the log home. But there was enough snow that the stack of logs needed to be dug out of the snow, if they were there, and it was time-consuming. To top it off, the logs were no closer than 20 meters away on the other side of a ditch. So I spent two or three hours finding and grubbing out logs, cutting them in manageable lengths, and then dragging them to the one-ton truck, which was patiently waiting for me to load it up. Some of the wood was not good, but it was manageable. This sort of a motley pile of wood my dear wife and children made do with. The wood that was wet or green was laid on the stove or around it and left to dry out until it could then be put in the firebox.

So among all the tasks that we had in our daily schedule, cutting firewood was important, if marginal. Our purposes in moving to the Peace River Country--to get away from commercialism, greed, nonrenewable practices and non-sustainable lifestyles--were closest to being realized with our heating and cooking methods. Wood cutting was part of that. But even then, I thought that the dependency on gasoline for the chainsaw, the tractor and the truck were compromises with our purposes, and only after a settled time could we cut wood by hand, or better, gather wood as the Indians did, which would then be sustainable.

I remember my grandfather, Opa, living in northern Wisconsin heated entirely with wood he cut with hand saws. The large blocks were mostly cut into blocks with a two-man saw. This saw was just as fast as a chainsaw, which has become popular and ubiquitous in the last 40 years or so. The two-man saw needed two people to keep it going, one on each end, and this was its weakness. Heavens, one person ought to be enough to cut wood, which is what I mostly did with my chain-saw. But having used a two-man saw, I know it is fast, safe and quiet. It also uses no gas. So Opa would use the summer to cut enough wood for the long Sawyer County, Wisconsin, winters. Mostly this was done after the trees were cut into blocks or the smaller diameter limbs into lengths. The blocks were split with a maul and wedges by hand. The lengths were cut with a buck saw, and I too had one of these. It worked well but did require time, and that was my restriction: not enough time. So while I used a buzzsaw with the John Deere B, Opa used a buck saw and a saw buck. The buck saw was a sharp blade, a saw, fastened at both ends with a 50 centimeters lever that was spaced, fastened to, another rod at the middle. At the top was a small cable with a turnbuckle fastened to the two levers and used to tighten the saw blade. Opa spent a good amount of time sawing the wood, and when I was there over summers I often grabbed the saw and would cut some of it. It worked, and I have sketched the view of a typical buck saw.

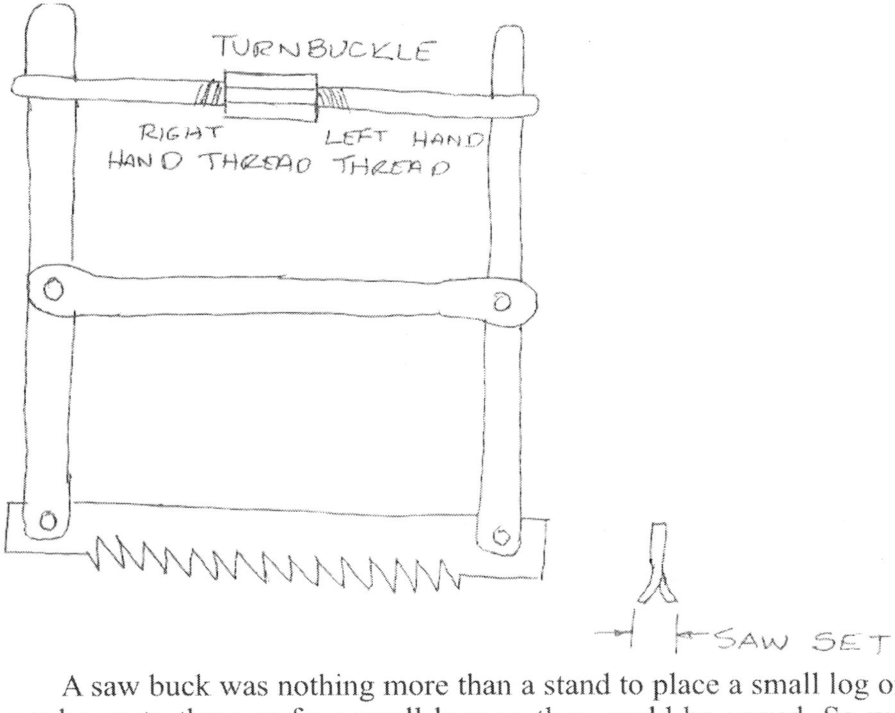

A saw buck was nothing more than a stand to place a small log or maybe up to three or four small logs so they could be sawed. So we built a couple of saw bucks using some scrap lumber.

The use of a buck saw and two-man saws was more gently closer to a sustainable practice, but even it was non-sustainable. Finally, heating and cooking needed to be done with a sustainable wood supply and gathering dead sticks and tree limbs, carrying them to the camp or home, and breaking them. It would be foolish to expect today's citizens to behave this way, but eventually it will have to come. Even the kind old cook stove and the pot belly stove were not, in the long run, sustainable. A fireplace, not the insert type but a real fireplace, is approaching sustainable methods, but it too requires some technology. Cement, mortar and fire bricks are nice but eventually not sustainable. A campfire or a small circular fire in a teepee, a wigwam, or a lodge is sustainable. The free lunches will play out, and people will be relegated to either sustainable behavior or extinction.

Here in Canada we had uprooted ourselves from the traditional social and technological culture, and compromises needed to made early on. Would I get to a point where the wood could be cut by hand or better yet gathered? So it would be a race, so to speak, to see if our long-term plans could be realized.

CHAPTER ELEVEN
January Cold

The first winter in Canada became a challenge for us all. The winter of 1978-79 turned out to be a record cold one. The natives kept telling us that it would get colder, beginning almost from the moment we arrived in Canada in late August 1978. By October, when it was freezing almost every night, the temperatures just kept creeping downward.

During a cold day in October we were visited by a fish monger. He had a pickup truck of some sort with a big camper shell and was selling fresh salmon from the Great Slave Lake, some 400 kilometers north of us. It was interesting because in late October 1978 a Russian satellite or missile with a nuclear cargo had crashed into the Great Slave Lake. It caused quite an international furor, and everyone was pointing fingers at each other, but the Great Slave Lake now had a Russian warhead in its depths. The salesman assured us that the fish were not contaminated by any radiation or whatever. We went ahead and bought some of the fish even though we knew he had no idea of its safety. They looked good, and we wanted some fish for a change. But it was another haunting event that told us that we had moved over 5,000 kilometers to get away from civilization or at least the worst of it; there is no way to do that.

I remember well in middle October when we were living in the Tomzak place that one of our good neighbors, Nick Kanda, came over one evening just to see how we were doing and in his clear Canadian speech said, "It's going to get cool tonight, eh? Maybe –20 degrees C or so."

I thought to myself, "Cool? It has been cold by Ohio and Indiana standards now for a month, and Nick says it is going to be cool at -20 C." I told him, "It has already been -15 C, and that is pretty cold already, isn't it?"

"Around here cold begins at -40 degrees C, and from there on it is a matter of just trying to keep from freezing up." For some reason, knowing Nick for just maybe two months or so, I had the distinct feeling that he was telling the clear truth and not trying to impress a foreigner with local bragging rights.

So the temperatures began spiraling downward. It was a gradual drift downward, but by early January the temperatures did not ever reach -18 degrees C (zero degrees F). And by early evening, with the short daylight hours and snow on the ground continuously now, it often drifted down to -40 C. This is a nice talking temperature for the non-technically inclined because at -40 the number is the same in C or F, that is -40 degrees C is -40 degrees F. From that point on downward, the minus C temperatures are lower for the same coldness. So at -60 degrees F the Celsius temperature is just at -51 degrees C, which happens to be the temperature that we would feel before February would come.

The neighbors were well aware of what was coming, and they were mostly prepared for it. Beginning in early January, when we had just been in our log home only a month or so, we were getting accustomed to working just to keep the home warm and prepare food for the children. Joy was nursing Sunshine and needed the best food for herself and the baby. It was not fancy food, but basic and solid: cooked wheat or barley cereal, potatoes, carrots, turnips, all vegetables from the garden that the Peacheys had planted. We were constantly reminded of their helpfulness.

We even had a portion of a moose shot by a local, and a young yearling bear became part of our diet as well when a neighbor, Ken McRann, saw the bear in a field and shot it. I was always outspoken about how hunting should not be done, but at this time, in the cold of winter, with no well-prepared larder of food, I needed to be more humble and even meek.

Our log home, which we occupied by the first of December, had a pitcher pump installed at the kitchen sink looking out to the east over the old dugout. It was a picturesque setting and was one of the high points of our Canadian adventure. The home was perhaps 20 meters from the shore of the dugout, and we have pictures of the home taken from out on the dugout during January. It was indeed beautiful and romantic. The pump was purchased in Ohio and was the old reliable design, a piston with a leather cup that allowed the piston to lift the water and a spout directed the water into the sink, a wash tub, or whatever. It was a good arrangement, and the water came from the dugout. Remember that by this time there was maybe 35 centimeters of ice covering the dugout but our water was pure and clear, not like the water from our Tomzak place. The key here was that our dugout was very deep and there was a large amount of water under its ice. So we were set for the winter in terms of fresh water. Most of our neighbors, even the Kandas, had small dugouts, and the water would become rank by March or early April with bugs and algae. Some neighbors melted snow in tubs and used that for drinking and cooking, and others even went to Dawson Creek and bought water. But while we didn't design it this way, we really had a good water system. The waterline from the dugout was a polyvinyl chloride (PVC) black pipe with an inside diameter of 2.5 centimeters, enough to supply us with probably a gallon a minute if one pumped continuously. The line had been installed in late October or early November and was buried maybe 2 meters deep. The line had a one-way valve at the end, or the end in the dugout, called a foot valve. This valve assured that there would be water in the line at all times, once we had primed it, and prevented backflow of water from the house back into the dugout. The foot valve was a crucial part of the system, and since it was at the bottom of the dugout, maybe 15 meters under the surface, there was no way to check it once it was dropped into the dugout. So at no point in the waterline from the dugout to the house was the line less than 2 meters under the ground. The water should not freeze up, even if it was exceptionally cold.

Rules of thumb are always a quick way to determine how deep to bury a waterline to prevent the water from freezing in the winter. It varies from location to location and what kind of ground the line is buried under. For instance, in Ohio if one buried a waterline 3 feet, or 1 meter under the surface, the water would not freeze. There could be exceptionally long and cold periods, but unless the ground above the line was compacted by say a pavement or parking lot, 3 feet deep would suffice. In the Peace River Country of upper Alberta and British Columbia, one needed to go down 2 meters since the weather was much colder and for a longer time period. All the neighbors who had previous experience with burying waterlines told us this was good.

As with everything we humans do, there turned out to be one problem (probably more than one, but this was the pressing problem). We buried the line in late October when the temperatures were drifting down, well below freezing at night, and this figuratively soaked the ground under the house and everywhere else with cold. We had a frozen mass of ground down about a meter and a half or more. The waterline came under our house foundation, buried to 2 meters deep, and at the kitchen cupboard it need to rise up and connect to the pitcher pump. In the following figure I have sketched the configuration of the arrangement of the waterline at the kitchen cupboard.

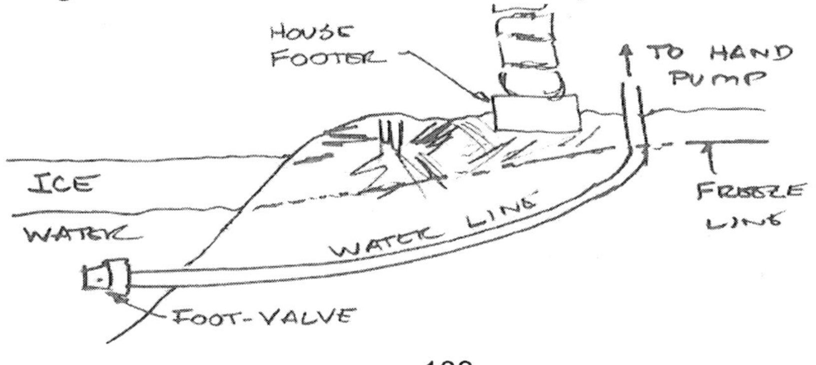

As is clear, I hope, the waterline passed through the ground that had become thoroughly soaked with cold since October. There was just no way for the waterline and the water to avoid this path. While one was pumping water, the water would flow from the dugout at maybe 10 degrees C through the waterline which was about 40 meters from the bottom of the dugout to the kitchen. If the water would be pumped through the line, it would reach the pump at maybe 5 degrees C since it would be cooled by the ground. In addition, during the course of the day, with both stoves going, the family active in the house, and with water being pumped for cooking, washing and drinking, the ground immediately around the waterline under the house would be warmed a bit by the water flowing through the line.

But at night, when the January temperatures were getting to -20, -30. -40, and even -50 degrees C and even though the ground around the line may have warmed up a bit during the day, the large mass of cold ground brought the waterline temperature in the one and a half meters just under the kitchen to below freezing. With the water in the line just sitting there, the water would drop below freezing in the line so that in the morning the pump could not supply water to the house; it was ice.

The first time this happened in early January I became very discouraged. Not only were we struggling to keep warm and have enough food for everyone, but now we had no water. There would be no practical way to thaw the water in the line with heat, even if we had a blow torch or whatever. So heating the line was out. Then I thought of a method that I felt would work.

First I disconnected the waterline at the pump. There was ice at the outlet, which I already knew. Then I poured regular table salt on top of the ice in the waterline. This thawed the upper surface of the ice, and I needed a way to get that saltwater out. We had lots of straws, the kind our children liked to use to drink their milk, soda or water. So I took a straw and sucked the salt water out of the waterline. It worked slick, and within a minute I had sucked out the melted icewater. But I had only removed maybe 1 centimeter of the ice in the line, so I went ahead and poured some more salt in the line. Again, using the straw I was able to literally sink the straw down another 2 centimeters or so. By sucking the saltwater out, a turbulent action was created which helped to melt the ice with the salt. After maybe five minutes I had removed the ice down to the length of the straw. So, I added more salt to the line, now maybe 12 centimeters down. Then I spliced another straw to the end of the first straw, so I now had a double-length straw. This procedure worked down until I reached liquid water, and I then reconnected the waterline to the pitcher pump and we were back in business. I felt good about all this and assumed now we had reliable water.

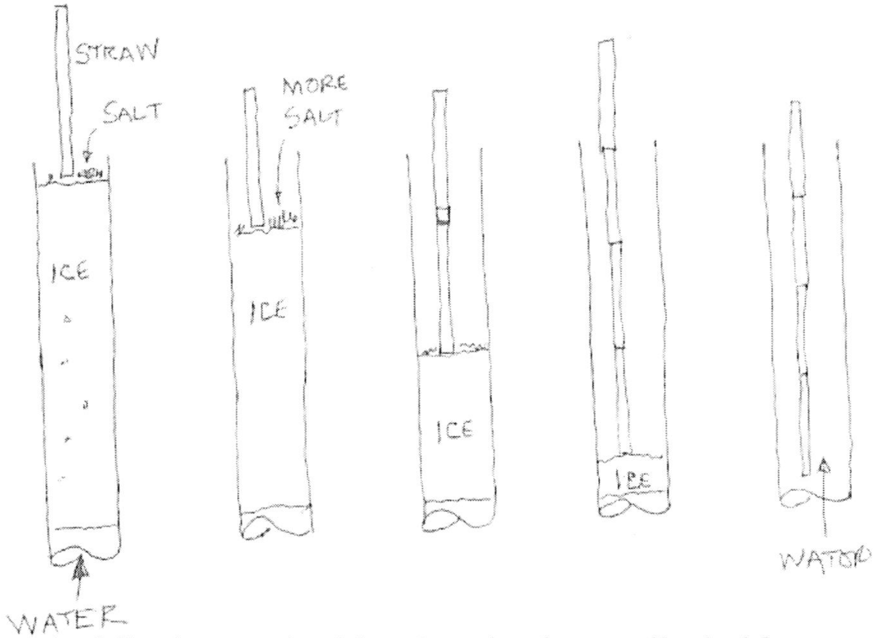

The following morning I found out that the waterline had frozen up again and I had to repeat the procedure, and this became a daily routine of spending 10 to 15 minutes thawing the waterline. As this continued, I noticed that each morning I had to go down a bit farther so that eventually I had to have three straws and then four straws. I did a mental extrapolation and figured that by the end of January I would need six or seven straws spliced together to suck out the saltwater and open the waterline. The phenomena of this ice freezing lower and lower was due not only to the January temperatures getting lower and lower, but the fact that liquid water conducts heat better than ice and in a rather complex way froze water deeper in the line.

As it turned out, we reached an equilibrium at four straws or about one half a meter down in the waterline. I made sure that I was up early in the morning to get the waterline thawed out, just to be sure, because I was also sure that four straws spliced together was the maximum number I could do. My suction was reaching a maximum, and a fifth straw would have exceeded my suction capabilities. So it all worked out in an interesting way. Just as a help in visualizing what transpired in the waterline thawing operation, I have sketched it in the following figure. And, during the following winter of 1979-80 the waterline did not ever freeze up because the house was occupied during all the fall and during freeze ups in late September and October. This meant that the ground under the house never became less than say 5 C, enough to not freeze water.

But even with the pitcher pump now furnishing reliable water, we were finding that January was relentless in becoming colder and colder. Our three oldest children, Kurt, Lori and Tim, were going to school at the Baytree Mennonite School, and we had to take them there. The school was located on the Peachey farm, about 20 kilometers west of our home, and we were finding it hard to get our vehicles started. The two tractors, the JD B and the JD A, were crank-start machines, and they seemed to be able to run regardless of the temperature. They were stiff and hard to start, but they did start. The Chevy C10 pickup truck was the vehicle of choice for getting the children to school, but at temperatures below -30 C, everything was going beyond its limits. Cold reduces the cranking power of a battery, cold increases the friction and stiffness of the engine, and the cold inhibited the gasoline's vaporization needed to fire the engine. And the temperatures seemed to just keep getting lower and lower. At first I would run the truck for an hour or so just before going to bed, at maybe 10 o'clock. Then I would wake up at 2 o'clock and go out and start the engine and let it run for an hour or so. Even this, with only four hours of cooling off, the poor old truck would groan and finally fire up. Waiting another hour probably would have meant not getting the truck started. Running the engine an hour and then getting up early, at 6:30 or so, gave us the transportation to get the children to school. But this whole

operation seemed to be counter to what we had intended when we moved to Canada. Here we were trying to keep a gasoline engine running at extremely low temperatures, and this surely was not sustainable. And to top it all off, it was not fun to wake up at 2 a.m. and spend an hour just so the truck would start in the early morning. Life was not getting simpler at all.

I did try putting a pan of glowing coals under the engine at night to keep the engine warmer, and this did work somewhat, but I felt very uncomfortable, and when the temperatures did drop to below -40 C, we finally told the teacher that we just had to skip those days or come in late.

In the third week of January we woke up one morning, and it felt exceptionally still and cold. The thermometer indicated -60 degrees F, or -51 degrees C, and one could just feel the cold in the stillness of the air, the hazy sunrise, and the pronounced crunch of the frozen ground and snow cover. It was an experience that made the total journey worthwhile but also brought up a number of heretofore unacknowledged problems. Our two horses were left outside, and I would go out, after thawing the waterline, and make sure they had hay and some oats. I occasionally treated them to a bucket of water which they thoroughly enjoyed, but horses can get their moisture requirements from eating snow.

Now, with -60 degrees F temperatures I decided that they needed some help. To make the decision obvious, a few days before, on February 10, Morning Star was lying on the ground and my initial reaction was one of panic. Horses don't like to lie on the ground, particularly if it is very cold. So I gave her my undivided attention and covered her up with two horse blankets. I rubbed her to get her activated, and that seemed to help her. I gave her some ground oats and barley. She seemed to like that, and that reassured me that she would be OK. She then stirred and got up. I gave her some warm water, and that hit the spot for her. And this seemed to be a forewarning of getting some shelter for the horses and a cow I intended to get. We had the log granary that had a nice wooden floor and was big enough for three good-sized animals. It even was partitioned off so the two horses had their own stall, and that was where I would keep them for the duration of the cold weather. During the day I would let them out to get some sunshine and fresh air.

The hay I was feeding the horses came from the neighbors across the road, Fred and Irene Pilsworth, and it was good to have. I got some barley and oats from good Nick Kanda, so the horses were set for the winter.

The first two weeks of January we stayed home mostly. As I said, after attempting to keep our vehicles running and not doing that very well, we did manage to get the children to school maybe half of the days, and then we would check our mail at Baytree. So our encounters with neighbors and friends were infrequent. But the third week of January we needed to get to town for supplies and just try for a break. At Dawson Creek, particularly at the Co-op, which was a general store with a grocery section, a hardware section, a clothing section, and even a restaurant, the main talk was the weather. And it was obvious that even the natives now felt the cold. With two days of -60 degrees F and continuous cold weather, the conversations among the natives centered just around this. The snowfall was not exceptional, although with very cold weather any snow was bound to stay a good period. But at least there were no blizzards and the wind actually was almost nonexistent in the very cold weather. Snow would pile up on fence posts, power line poles, and any place where it couldn't slide off. A meter of snow on the ground meant that a fence post could have 50 centimeters piled on it. It was not wet snow, so a good wind could blow it off, but the daytime sun, whenever it occurred, would melt small amounts of the snow, and then the night cold would make a crust on the top of the snow. So powdery snow eventually, after a few days, became crusted frozen snow. It was a beautiful sight to behold.

So being in Dawson Creek turned into a treat for us all, and we would try to get something for the days at home: some special dry cereals to cut the normal breakfasts of cooked oatmeal or wheat or Red River. We were not into soda pop, but occasionally even a six pack of Canada Dry Ginger Ale or root beer was purchased. We did get sugar, flour, salt and baking powder. Raisins also were a good addition to the homemade cookies. And Postum, tea and honey always were in the bundles to take home, if we were running low of them. We did also sometimes get some "store bought" bakery goods or ice cream, although we often made ice cream with our hand-crank machine at home. Clothing for the children was always checked, but we were mostly still using hand-me-downs with neighbors helping to keep the children covered.

The month of January 1979 became a rather long one physically and psychologically. It remained cold the full month, even though there was always the promise of a Chinook when the temperatures could zoom to unnaturally highs, 20 and 30 degrees C. But this month was consistent in its coldness, and by January 31 we had had, literally 31 days of temperatures that did not reach above zero Fahrenheit, -18 degrees C. It was reported by the neighbors that this was the coldest January on record, and we were sure not disagreeing. But on the last day of January, January 31, 1979, the temperatures skyrocketed to the mid 40s and snow dripped off the roof of our house for the first time in the year. It was a welcome sign. And we survived it without electricity, a telephone or reliable vehicles.

CHAPTER TWELVE
Hints of Spring

The Canadian weather had given us a true welcome, and somehow we survived and weathered the cold January temperatures. The snow was piling up too but not as much as we had expected. There was probably 1 meter of snow on the ground, and since at that time winds were not a major problem, the drifts were not more than another meter. I remember one of our neighbors, Arnold Hayward, did the snow plowing and one evening I heard this beep, beep, beep and saw blinking lights on the road. I couldn't figure out what was going on. Later I learned that the snow plow, a road grader, was equipped with a horn that would beep whenever the grader was put in reverse. Since that time many trucks, off-road vehicles and even Sport Utility Vehicles (SUVs) have these beepers to warn of the fact that the vehicle is backing up, for what that is worth.

I remember on the road leading north out of Baytree, a north-south road and one which would experience winds crossing it, there was a snowdrift of about a meter and a half, but even in our front-wheel-drive trucks we managed to get through. We did have occasion to go to Baytree as that was our address and they had a Post Office box in the store where we picked up our mail. The store reminded me very much of the later general store that was built in Lemington, Wisconsin. While the details are not the same, both stores were white frame structures with a gable roof and a front facade to give the appearance of a "square building."

The Lemington store had a single gas pump in front, one that required the owner, Ted Schlau, to reciprocate a meter-long rod back and forth, pumping gas from a tank in the ground to a glass cylinder 2 meters off the ground. When the gas level in the cylinder reached the amount to be purchased, I believe it held five gallons, that gas could then be dispensed into the vehicle, car or truck or whatever, gas tank. If someone needed more than five gallons of fuel, the process needed to be repeated until the full amount of fuel was pumped out. The Baytree store was a general store without the gas pump in front. On the left of the store was a log town hall, and on the right a house all three in a row, on a road that angled off Highway 49. It was originally Highway 49, but just a few years before it was bypassed by a new highway. Across the road from the store was a United Farmers oil, gas and diesel station. It was not an actual gas station but did dispense fuel for the local farmers. We bought our gas there too, purple gas. Canada tinted gas purple that was dispensed to farmers to distinguish it from the regular gas, petrol, for city folks and non-agricultural purposes. That was another 20 cents more per liter. And at that time the gas still was leaded, although Canada would during the time we lived there implement the de-leading of their gas. While we were there, we could get leaded fuel or unleaded fuel. Canada seemed to be more environmentally alert, caring of their citizens, and willing to do the right thing than the United States. That was another reason I felt good about moving north. In the US everything finally came down to whether the fat rich businesses and their stodgy politicians wanted to do something. I remember that Canada implemented the metric system, or really the Systeme Internationale (SI), and they did away with the clumsy "English" system of units. Pounds, feet, miles, acres, Fahrenheit and gallons would be no more in Canada. And it all went smoothly, and the people now have a reasonable way to measure and compute. In the US, when we left, the conversion was proceeding, but the obstructionists, the dummies, and the egocentric business people were making noises that they didn't want to convert to the SI system; it would cost too much. It turned out that two years later, under a new President, that is exactly what happened. The SI system was dumped in the US, and we still have this ridiculous Fahrenheit, foot, mile, gallon system. It probably fits the country.

So we bought gasoline in litres, drove 40 kilometers to Dawson Creek, and our thermometers read Celsius. And zero Celsius was just freezing, while 100 Celsius was boiling water. Oh, the Canadians were obstructionist as well. Some of the people put leaded gas in their vehicles even when it should have had unleaded. The newer vehicles had catalytic converters that essentially burnt the unburnt fuel in the exhaust gases by a catalytic process. Lead destroyed the catalyst, and the converter would not function then. So the vehicle would emit the unburnt fuel into the atmosphere. It was really childish behavior, and it was going on in the US as well, except that it was deadly serious and a waste of natural resources and inefficient. I even now cannot understand why anyone allows for inefficient combustion processes, dumping process wastes into the rivers and lakes and the air. These are supposed to be rational intelligent beings, but my experience seems to indicate that the only thing most people want, including the business types, the accountants and the efficiency experts, is a quick easy solution that doesn't require them to actually work or think but that makes them lots of money.

Baytree was just a quiet stopover, not even a hamlet. But I remember going to get the mail and buying some groceries, what they had, some ice cream or other treats like candy bars and soda, and talking with the proprietor, Mr. Budde. He had a wife and two children, and they had moved down from Yellowknife, Northwest Territory, about five years before. They lived along Highway 49 about a kilometer west of the store on the south side of the road. He was friendly and willing to expound. Particularly, he told of how much colder it was in Yellowknife than in Baytree. I am sure he was not mistaken since Yellowknife is about 200 kilometers or so south of the Arctic Circle at a latitude of 63 degrees north and it was on the north shore of the Great Slave Lake at the west end. It must be a rather bleak sort of town above the tree line, a real frontier settlement, and the winters were surely cold. While around Baytree it was common to plug an electric heater into the engine to either heat the engine block water or heat the engine oil, in Yellowknife one needed to also keep the battery heated or it would freeze up and be destroyed. And here in Baytree we always layered our clothing when we prepared to go outside--at least two pair of socks and well-insulated boots, a pair of long underwear, top and bottom over regular underwear plus one or two pants and shirts, with a good long insulated coat. A pair of regular gloves and over them a pair of mittens would be good, as would a ski cap to cover everything except openings for the eyes and mouth. Mobility would be reduced significantly, but one could survive outside well. In Yellowknife, apparently, one needed an additional layer all the way around to remain warm enough. But whether in Baytree or Yellowknife, if one was active enough, one could remain warm with lesser amounts of clothing. But it meant keeping active and moving. I felt that Baytree was far enough north right now.

The weather remained cold, drifting down t o -60 degrees F a couple of nights and early morning and then warming up to -20s during the middays. On February 14, Valentine's Day, it was -60 degrees F at 7 a.m., cold and dark, although the daybreak was promising. Here in the northern latitudes the daybreaks were two hours or more, and similarly, the sundowns were another two hours or more. One had to get used to that phenomenon as on clear days and nights darkness was a bit relative. At 7 a.m. it was hinting of daylight, and by noon we had a warm -20 degrees F. We had just completed a solid week of cloudy and snowy weather leaving us with more than a meter of snow on the ground. So we had a nice stretch of sunny weather, if not warm. With the clear days, we also had clear nights and often we would go out after sunset, maybe even around midnight, and watch the stars in the clear sky. Up here there were no faraway surface lights, like a major city or even an oil rig campground. And the silence was comforting and stimulating for thinking about creation or whatever. Even the northern lights, the Aurora Borealis, would make a good show. The night of February 24 was one where there was a quarter moon, just a sliver, and suddenly, around 11:30 p.m., the northern lights began to streak across the sky. They moved in profound and unpredictable patterns, going from the north to the south and west to east. They were even colored with mild shades of pinks and greens and light blues and dim oranges. They would even crackle and pop sometimes; better than a movie show. And even with the cold air and cold sky, if one were bundled up and alert, it was comfortable. I suspect that no one has frozen to death watching the northern lights. Again, the keyword here is watching, and sometimes we drift off and physically we are out in the night air but not watching but inert. That could be a deadly tone.

The publisher of my textbook wanted a revision to extend its published life, so I spent many an hour from late December to middle February rereading and revising the text. I only had a mechanical typewriter, but that served well enough to get it prepared, and one satisfying day in late February I took it to Baytree and shipped that package off to the publisher. I had been receiving royalties every six months since the book was on the market, and that gave me a bit of a financial income. It was not enough to keep us going, but it helped. In addition, the work I had done to earn the Doctor of Philosophy degree in high pressure physics was done under a grant from the US Air Force. They required a final report on the work, even though I had written and published the dissertation. My advisor asked me to help with the writing and also do some final calculations that would satisfy the grant, so that too was a bad weather project for me. I remember doing much of the work for the book and the grant final report in the south upstairs bedroom. There was a nice dormer with a window overlooking the dugout, small lake, and it was warm. The conditions were quite acceptable and inviting, and it all went well. I received some financial payments for the final report work too. And that documentation was shipped out of the Baytree post office as well. I do remember that since we had no telephone, communication with the book publisher and my advisor was mostly done by mail. Our neighbors across the road, Fred and Irene Pilsworth, had a telephone, and I would go there often to have teleconferences. I felt a bit imposing on them for doing that, but I believe they enjoyed having me around and it was a different experience for them. It wasn't ever that a PhD in engineering was visiting them or living across the road. So my umbilical cord with my previous activities in Dayton, Ohio, was not completely severed, and we had just a trickle of income. We also were getting a family allowance. Actually it was a national Canadian government policy to distribute a monthly check to families having children. Joy applied at Spirit River, Alberta, for the funding. It was a policy of giving something like $30 per month for each child, with a stipulation that you needed to have a low enough income. We qualified easily for that, and with 6 children we received around $180 per month allowance. This sort of thing was unheard of in the US. Again, another case of welfare for the rich in the US and welfare for the poor in Canada. The US seems to like Trickle Down policies (piss on them), and in Canada a bit of fairness was applied.

On February 28 we suddenly found it only -15 degrees F at sunrise after being in the -30s the day before, and it was still sunny. By noon it hit freezing, 0 degrees C or 32 degrees F, and it felt balmy. By March 3rd it was near 50 degrees F in the early afternoon. We were having a Chinook, and that was always the promise held out by nature around here that kept everyone going, I think. On March 4 Sunshine was three months old and growing like the wild oats in the Peace River Country in July. She was learning too by leaps and bounds, words forming and actions of all kinds. And on March 8 we celebrated our oldest son, Kurt's, seventeenth birthday He was getting mature and just a joy for us all. It reached 50 degrees F in the afternoon, and we just relaxed. Even the waterline from the dugout to the kitchen thawed out without my efforts, and it was one more of the high points since we had moved away from Ohio. Interesting also was that on March 8, 1962, Kurt was born in Richmond, Indiana, which was a raw cold day with windy and snowy conditions, not Chinook weather at all. But we had a perfect son.

On March 10 the Chinook left us, and we were again back to the regular weather. We spent the day gathering and cutting fence posts for our pasture that we would put in when the weather permitted. And even with cooler weather, on March 14 we saw three Pine Grosbeaks around the house. The dugout was frozen over, but we sprinkled some oats and barley for them to feed on. I suspected they needed some water, so we did put out a pan for them. It was kind of touch and go since our two dogs were making their rounds on the farm, and we didn't want them to have a meal of Pine Grosbeaks. We determined that the birds were indeed Pine Grosbeaks from our Birds of Alberta book which had a description of them as well as a sketch. They were unmistakable with the male having spectacular dark orange top feathers, alternating dark brown and white wing feathers, and a proud brown tail. The female was not as conspicuous in color, as is often the case in wild fowl. It seems that the Pine Grosbeak wintered in the Rocky Mountains and would come to the northern plains when the weather moderated. This may have been an omen of the coming of spring in the Peace River.

CHAPTER THIRTEEN
The Cow

With six children and trying to be self sufficient, the need for a cow seemed obvious. When we lived in Ohio on the farm, we had at least one milk cow for the last five or six years. We milked the cow, and I recall even supplying raw milk to some of my friends at the University of Dayton. We often made homemade ice cream, and Joy was able to make cottage cheese and yogurt when she had time. And we had butter churns or even a quart jar to churn butter. So our children were all used to a cow and liked milk, butter and cheese. In addition we now had six children, and while we were able to purchase milk from neighbors, it was not the same as having our own cow.

A cow was a necessity, and we bought a half Jersey/half Hereford cow and a one-week-old calf from Willard Stutzman on March 22, 1978, for $600. We had spoken for her back in October 1978, and Willard told us then that we should wait until the cow had a calf and would freshen, that is, begin to milk again. This turned out to be a good arrangement as the cold January would have given us even more hardships with trying take care of a pregnant cow in the bitter cold. The Chinook in March affected our lane, which up until now had been frozen solid and snow-covered. Now it was rutted mud, and this mud was not the clay in Ohio. Somehow it had a tenacious ability to stick to everything, and yet it was viscous so that we would slip and slide. We did make it to the barn area, but this was a premonition of what would be in store for us, not only on our lane but the highway, when the serious spring thaw would come.

The cow, Flossie, was a nice mild-mannered animal. The Jersey could be seen in her with her dark eyes and long eyelashes. And she had the build of a Jersey as well: delicate and small. But she also was a bit stout and seemed to have the ruggedness of a Hereford. Willard told us that she was a good milker, and the calf, whose father was a Charloix, would be a good calf. It took only a few days to realize that Flossie was not giving enough milk, and since the calf was not weaned yet, the milk that we did get went to the calf. The first sign of trouble was that Flossie just wouldn't eat anything, and even drinking water was painful for her. But this too became insufficient for the calf, and in all my experiences with cows, I had a bad feeling that the cow was not well. I kept thinking that Willard had some reluctance about selling Flossie to us, and I knew he didn't purposely sell us a sick cow. He had made us the offer in early January, but when the time came for the cow to calf, he wanted to back out of the deal. Finally, I think he realized that his word needed to be solid, and he did sell us Flossie. But she was not doing well, and in addition we were losing the calf. The calf had no margin at all, and finally she died. I couldn't think what the poor cow had that kept her from giving milk. I talked with Willard, Jesse Peachey and a number of the neighbors around us, and it finally became obvious that if we didn't do something quickly Flossie too would be lost. The nearest veterinarian was in Fairview, Alberta, and it was about 120 kilometers (75 miles) away. We called the vet, and he thought, without seeing the cow, that maybe her stomach was twisted around. This seems to be something that happens with cows; during birthing the stomach gets tangled up and twisted around the intestines, causing the cow to not be able to digest food or even pass water. He said that if I brought her in he would do what he could. So I didn't really mull the decision over long; I loaded Flossie in the racks of the one-ton truck and headed for Fairview. It was a cold morning at the end of March, maybe 10 a.m., but the days were along short and dark in March, so I kept some water and feed in the rack in case Flossie decided to eat or drink. I went by myself because everyone had school, Joy had to take care of Sunshine, and Kurt and Lori needed to be around to help out and keep the stoves going. The drive to Fairview seemed to take forever. I wanted to get Flossie to the veterinarian to relieve her of obvious pain and hunger and thirst. I drove 16 kilometers through Gordondale on Highway 49 and on to Spirit River. Highway 49 in the spring of 1979 was a dirt road. It had been regraded with wide shoulders and new ditches and built up above the countryside in the fall, but it was dirt. The following summer it would get a pavement. Fortunately it was cold and frozen, so the road was hard. So I was able to drive maybe 70 kmph (45 mph), and I kept checking in the rearview mirror to see if Flossie was still with me. She was, but she was lying down. It was another 48 kilometers to Spirit River, which I had been to a few times. Spirit River was, for us, the nearest Provincial seat. The regional court was here, the RCMP, and the other bureaucratic offices, and it was the largest town east of Dawson Creek at maybe 1,200 citizens. Spirit River was quiet and didn't slow me down at all. Rycroft was another 12 kilometers down 49, and here the highway became paved. Also, the spur of the Northern Alberta Railroad that served Spirit River paralelled the highway. Rycroft was a small junction for the railroad, but other than a few grain elevators, it was very quiet. Highway 49 intersected Highway 2 at the east side of Rycroft, and here I turned left, north onto 2, headed for Fairview. Highway 2 was the main Alberta highway between Edmonton and Dawson Creek, but it meandered rather indirectly. It went north from Edmonton, turning west through Slave Lake, then turned north again to Peace River, the village, and from there turned west again to Fairview. It then turned south through Dunvegan, Rycroft and Grande Prairie before finally heading west to British Columbia and Dawson Creek. But I would only be on Highway 2 for about 45 kilometers. Heading north, with a reasonably good pavement and broad shoulders, I made it to Dunvegan, about 20 kilometers past Rycroft and then to Fairview. Dunvegan was a historical site, and as most people don't know, northern Canada was explored by Europeans in the middle 1700s. Dunvegan was a small fort and church. The French needed to placate their homelanders in France and tell them that they were trying to convert Indians to Catholicism. The raping of the land, flora and fauna, was not usually pointed out.

After about a two-and-a-half-hour journey Flossie and I arrived in Fairview. Fairview was maybe 1,500 people, and the vet clinic was not hard to find. Looking back I would be hard-pressed to find it, if it still exists. But I remember getting Flossie out of the stock racks. I think maybe the vet had an elevated loading dock, so we didn't have to put the chute to the ground level. We negotiated Flossie into the clinic and into a concrete-floored inspection room. It was very austere and had a drain in the center, mostly I am sure, for the water, body fluids and wastes to be flushed. The vet proceeded to look Flossie over very thoroughly; he seemed to know what he was doing, and I sure hoped he knew. I recall that he finally did an X-ray and verified that Flossie's stomach was twisted around the intestines. He said the quickest and least expensive or most economical fix would be to try to roll Flossie over and thereby righting the stomach. He said this method worked sometimes, but if that didn't work, then he had to operate and untangle the stomach. I can still see poor Flossie put on her side; the vet seemed to know which side to put her on. We together rolled her over onto her other side. Seeing a cow getting rolled over was something I never thought I would see, and Flossie was in obvious discomfort and probably embarrassed. The vet looked at Flossie again and saw that the first turn over didn't fix the stomach, so we did it again, first getting her up and then putting her on her side. Another time we rolled her over, and again it didn't fix Flossie. At this turn of events the vet told me that he would have to operate on Flossie to get her back to normal. In fact she would probably return to milking normally, and we only would not have the calf. I left Flossie with the vet and would return in two days to pick her up.

I do remember calling the vet on the first day, and he said that Flossie went through the operation well and was on the mend. I could pick her up the following day, which I did.

We tried to pamper Flossie as much as we could. She was mostly kept in the log granary with the two horses, each having a stall to themselves. And with three large animals in the log granary, it was actually rather cozy. Flossie did return to milking and gave probably a bucket of milk, maybe 3 gallons a day at two milkings. Flossie was a gentle cow and rather easy to milk. And the milk was rich, what with a Jersey heredity. Since she didn't get a lot of sunshine, Vitamin Dl or whatever, the milk produced a butter that was a bit whitish, not deep yellow like that of summer milking would produce. She mostly ate hay and a couple of scoops of barley and oats. I often thought it would be nice to give her some ground corn, but corn was not known in northern Canada. Barley is probably just as good as corn, and she did well. We kept Flossie for our time in Canada and in late spring 1980 she had another calf. This time the calving went well with no twisted stomach, and the calf was healthy.

CHAPTER FOURTEEN
Spring Thaw

We were settling in rather well, considering that the activity was nonstop since the move. We had a reliable cow giving us milk, our two horses were content, and all we had to do was get enough firewood and keep the house clean. That was enough right now.

We were now approaching the equinox when the daylight hours would equal the nighttime hours, twelve. But it was still cold at night and dark when it was cloudy. One cloudy night, we had just finished supper and Joy and Lori were washing dishes when we heard a knock on the door. Who could it be, maybe the neighbors? I went to the door, and here were three men dressed for the weather with heavy coats and hats and seeming to be in a bit of distress. I asked them in, and they obliged. We gave them each a cup of hot Postum and asked if they had had supper. No, but they were not really hungry. In this country one didn't take that as a good answer; everyone could eat something at any time in the cold weather. So we had some bread and cheese and sausages and jelly. They proceeded to explain that they were coming down from around Peace River, the town north of us along Highway 2, and had car trouble. Actually it was a pickup truck, and even then pickup trucks were more common than cars. They noticed lights on in our house and wondered if they could stay overnight and call for help in the morning. I saw no reason to not let them stay, and Joy too was amiable to that. So we put them up for the night. We had a bench-like sofa that I had made that sat next to the big pot belly stove and at a window facing north. So one of them slept on that with some blankets. They did have some sleeping bags, and they went to the car to get them. They brought them back, and the other two men slept on the floor in the living room area. In the morning we had a nice hearty breakfast: eggs, bacon, and toast and butter. The toast was some of Joy's homemade bread, and the butter was Flossy's milk fresh-churned. One of the men remarked that the butter was mild, tasted good, but was not as yellow as that you bought in the store. Joy said it was *winter butter* and that we had made it. They all were interested and actually seemed amazed that winter butter was different than summer butter (?). They just kept remarking all the time that they really liked winter butter.

After breakfast one of them called into Dawson Creek on their Mobile (cell phone), up there pronounced Mo-Bile. Within an hour someone had come out with a service truck, and they all went into Dawson Creek. They left very appreciative of us, and we were all pleased as well.

Well, it could have dropped right there. I had no problem with strangers having trouble and lending them a helping hand. But that next Sunday at church one of the church members came up to me after the service and suggested that maybe I was being irresponsible in letting total strangers stay overnight in my house. I had endangered my family, four daughters, and two sons and my wife. And he also said there was a rumor going around that three convicts of the Provincial prison at Peace River had escaped and were last seen heading south in a beat-up pickup truck. And this was the topic of discussion for a week or more around the Baytree area.

Somehow it all just evaporated, and I even heard that someone had seen the three in Dawson Creek, although I don't know how they knew "the three" since they didn't see them at our home. I am continually amazed, or maybe discouraged is the better term, about people's behaviors and beliefs. I look back at that incidence and yes I was surely taking a chance and yes they could have been evil, brutish men who would harm us. Maybe they were the three escaped convicts. Remember that the word "convict" makes that person guilty already and all the bad things that go along with that. I think there are even some verses in the Bible that would support what I did. Look them up if you wish, but to me trusting a person is the first order of business we have with another person. If they show later on that they shouldn't have been trusted, that is their problem, not mine. Wouldn't it be a great world if our politicians and paranoid citizens would heed this?

We wouldn't need a War Department and weapons of "defense." We could actually love our enemies, and our hearts would be at peace.

The later days of March were typical of what we knew from farther south; every day could bring whatever: sunshine and warm, or sunshine and cold, cloudy and snow, or just cloudy. But the calendar told us that time was on our side now. Spring would come and then the nice long days of summer. We needed to do what we could while it was still frozen over. So, we went to Spirit River with Jess Peachey one day and got a small amount of timber. I was reluctant to do it, but the Province of Alberta, and all of Canada for that matter, encouraged timber cutting for lumber. And they let one cut on Provincial land, as long as you purchased a logging permit. They had a quota of, I believe, 20,000 board feet in one purchase. We purchased 2,000 board feet, I believe. I think they were still using old English units, so they didn't yet quote in board meters or whatever. But it cost us maybe $20 for the 2,000 board ft (bdd) of spruce. They didn't charge for the black popular or birch. That was "trash" wood and free for the cutting. So even here we kept finding the lack of understanding of natural resources and their wise use, mostly for just being in the woods. Once a tree is cut, it is dead and a hole is left, some wild animals have lost a home or shelter, and a human being has lost a spiritual friend. We would return to Spirit River the next winter to purchase our full quota, and we also had plans to put an addition on the house. And lumber could always be *used* or sold for cash.

On the last day of March, Kurt and I went into the Provincial forest south of Highway 49 in the slopes of the Saddle Hills. It was a characteristic Boreal forest with plenty of poplar trees, some sprinkling of birch and every once in awhile a dark grove of tall straight spruce trees. There were also a few pine trees, but, while these were preferred over spruce as lumber, there were not many of them. The spruce seemed to grow fast, so that a 20-year-old grove of spruce was imposing. The trees could be up to 10 meters tall and straight. They grew in clumps so there may be a hundred trees in a thick grove. They grew, and just naturally the lower branches would fall away and be bare. Up in the canopy, maybe starting at 5 meters from the forest floor, the limbs would be the characteristic tannenbaum, Christmas tree. They were quite impressive, and I just didn't want to cut any down. I often, that day and other days, would sit under the limbs of a clump of trees and feel the moss-covered ground under me, the spruce needles that had fallen and the sprinkling of the spruce or pine cones. There surely must have been squirrels, Eichhörnchen, living in the trees or at least stashing nuts away for the rainy days. The woods butchers, loggers and timber harvesters, couldn't care less. They probably would be pleased in destroying an animal's home.

But Kurt and I spent the day with Jess Peachey and also another of the church members, Paul Sadlier, cutting spruce logs. We cut what we thought were about enough logs to get 2,000 board feet of lumber after sawing, and I also cut a few black poplars, cottonwood trees. They just seemed to be a good tree and would make some nice lumber as well. And of course this activity was all on the honor system, and we were legal. Whatever that means. Jess had one of his tractors in a clearing where he had sometime before put his sawmill and where we could pile the logs for milling. And he was so helpful in that he skidded our logs to the clearing using chains and come-alongs and peaveys. We would come the next day to get our lumber from the logs.

On April 1 we spent the day milling lumber. Jess had cut his quota of logs for the equivalent of 20,000 board feet, and Paul had maybe 10,000. One board foot is a 1-inch-thick board, 1 foot wide by 1 foot long. It is really 144 cubic inches of wood, and 2,000 board feet was a good pile of lumber. If we had gone to the lumber yard in Dawson Creek it would have cost maybe $2,500, cut and planed. Our boards would be rough until we could arrange for planing, and that would only come the next year. But this lumber was good for structures and all sorts of projects. It was a bounty for us, and I just tried to forget the desecration that I had committed in the forest.

By now the days were getting longer and we had actually passed the equinox of March 21 when the daylight hours are equal to the dark hours. Our daylight was now longer than in Ohio or any place south of us. The sun was hanging higher in the sky during the day, although it never would be directly overhead or withering light. We had to make plans for the up-coming spring planting, fence building, and building constructions.

Charity turned five years old on April 14, and we surely had to celebrate that important event. We had a quiet party and enjoyed each other. As Joy was to write at the time,

Charity is now five years old. A very pretty and sweet little girl who loves her baby sister Sunshine.

Charity was not old enough to go to school, but occasionally she did go with her older sisters and brothers. She was a reliable help for Joy in taking care of Sunshine, and she demonstrated the best in human behavior: quiet and humble and caring.

Four days later we got our chickens from Dawson Creek. We had ordered them a month or so before, 50 pullets and 100 cockerills from a hatchery in Edmonton. The chicks were flown in to the airport, and that is where we picked them up. The cockerills were a hybrid cross that were supposed to grow fast for meat. The pullets, hens, where a cross between the Barred Rock and the Rhode Island Red. I always liked

Barred Rocks as they seemed to be the classic chickens; black and white striped. They were hardy and good layers. If need be they were good for meat but not as fast-growing as the hybrid meat chickens. And the Rhode Island Reds are known for their hardiness. Our pullets probably would be much better for nutrition than the hybrids, but that is another one of the dreary stories of human meddling with nature. So now we had 150 baby chickens, and while I had constructed a chicken coop, it was not heated and there was no fence around it. And we would surely have plenty of cold weather ahead. So we kept them in the hallway at the top of the stairway in the house. It had a nice dormer with the window looking out to the west, and it would give our chickens a safe, warm and sunny spot to grow. Over the night we got another half a meter of snow, and the temperatures remained in the -10 degrees C range, not warm.

And did they grow. We originally had a cardboard ring around them and a layer of newspapers on the floor. We fed them finely ground oats and barley and kept water there in a dish. And with the stoves heating almost continuously, the chickens were nice and warm. Within a week they had doubled in size and were getting lively too. A freshly hatched chicken is not quite a handful, but these were now a big handful. And we were running out of fresh paper to put on the floor. The feed and water were not a problem, but the space was getting cramped, and very soon the chickens found ways to get out of their pen and run around the house. So, it was becoming obvious that we needed to get the chickens in the coop. They were big enough and robust so they could handle a bit of cold. I built a fence around the coop with railing and even used some of the lumber we had milled. It would keep the chickens in and maybe keep wild critters out. We actually moved the chickens into their house on May 11, and that was a relief. The weather was beginning to act serious about warming up.

On May 22 we had sun and now the days were long. The sun came up around 5 a.m., and as I wrote earlier, that meant there were hints of the sun around 3:30 a.m., regular time. At sunrise, there was frost on the ground, but that would quickly burn off. It was now time for almost around-the-clock not having an excuse for working. Sleep would have to wait now until late September. We had brought along a nice grain drill, 4 meters wide. We had purchased it in Ohio at an auction sale, and it was in good condition. Grain drills are usually at the bottom of the priority list for farmers as they are only needed twice or so during the farming season. But I think this was narrow enough and was originally to be pulled by two horses that the conversion to tractors and wide machinery made the drill obsolete, even for a "*conservative*" farmer. And here was another of my hang-ups with the US and Canada too, mostly the sociology of the whole world. I always assumed using a word would be a good description of what it was used for. So to be *conservative* would rightly mean to conserve the earth's bounty, the natural resources, and the clean air and water. It would mean to conserve life and be against the death penalty, war, the war departments, and weapons and guns. It would mean to care for our neighbors and seek to help them. It would mean to not be gluttonous or grasping. Well, my experiences seem to indicate that those who call themselves *conservatives* are contrary to these descriptions. Anyway, some conservative farmer with a wide grain drill sold his stubby drill at an auction, and I was pleased with it. We had worked up about 20 acres the day before with our John Deere A and the pull-type disk. The ground was black, peaty and loamy here in the north field of our property. It had been fallowed for at least the past year, meaning that someone had just worked it with a disk or harrow to keep the weeds down and it was not planted to any crops. So it looked like as good a ground that I had experienced. I thought back about the Ohio ground, and this looked better; not so much clay and seemed to just call out to be planted to oats and wheat.

By the end of the day it had warmed to about 80 degrees F, and we seeded down the 20 acres of ground to oats. I also underseeded Alfalfa and clover for a hay field the following summer of 1980. The working up of the ground for seeding was straightforward. We had taken our pull-type disk with us, and it worked good. It was maybe a meter wide, not wide by the modern standards even then, but just the right size for us. Since the ground had been summer-fallowed the year before, the rich black loamy soil just churned up nicely, and it seemed to ask to be seeded down, almost sensually. And it was the common practice at the time in the Peace River Country that a field was summer-fallowed at least once every third year. That was a matter of running a disk or harrow over the ground and covering volunteer plants. The idea was that it would make the ground more fertile the following year. Farther south summer fallowing may not work good because of moisture loss in the middle or late summer. Here that was not a concern, and it seemed to work.

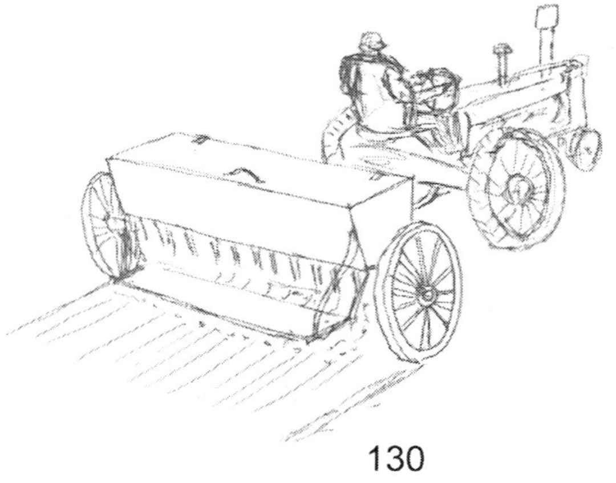

The next day we worked up another four acres and seeded it to wheat, spring wheat. So that gave us a bit of ground left in the 40-acre cleared land for making our summer garden.

Looking to the horizon in the west, where there was a muskeg, the poplars, birch and tamarack trees were showing signs of wanting to bring forth leaves or new needles in the case of the tamaracks. It had a promising blond yellow tinge and all around. In our woods, to the south of the seeded field, the willows too were starting to sprout their fuzzy nubs, and that gave a grayish tint to the scene. It was like a rhapsody or overture before the opera. Maybe Mozart, Beethoven, Wagner or Puccini should be here to observe nature's opera. And around Baytree and farther west, at higher elevations, many of the trees had experienced somewhat warmer weather, and they were a few days further along with leaves hanging on the limbs. It all was a marvel to behold, a true metamorphosis as Goethe observed many years before, if one was able to lift their head up and see, listen attentively to the earth's sounds, and smell carefully the scents of the awakening earth. It had been a long and persistent winter, but we were seeing another part of the Peace River Country.

The children were still going to school, but that was to end on May 25. The school children and their parents had a picnic, a program was presented by the students, and we all had a baseball game. Yes, even in Canada they played baseball, when it got too warm for hockey and broomball. It seemed like home in Ohio with the sunny day and the warm breeze and the spring signs. Even the birds were happy, and we did see some robins, if that is possible. Even they knew good country.

On May 27 the day started out cloudy and cold. The ground had now warmed up, and there was mud all over. But the hay fields and the road shoulders of wild clover and oats kept it all green. By noon the temperature had dropped to near freezing and a breeze brought a strange ominous sign. By 5 p.m. snow started falling, and in five more hours the ground had become white. All the new leaves and willow buds, and grass and clover shoots were now under a blanket of white snow. This wouldn't now last long, and a few days it was all back to spring, but nature here always wanted us know that variety is the power that it has and which people cannot control.

About this time it was becoming clear that washing clothes was a chore. We had brought with us the Amish washer, a tub with a swinging paddle that required someone to push a lever back and forth. I remember seeing one of the children or Joy operating it. After pushing the lever, and thus the paddle, back and forth maybe 50 times, the clothes were reasonably clean and could then be run between two rollers, the ringer, again hand-powered and that removed quite a bit of the water in the garments. They would drop into another tub of clear water for rinsing and again run through the ringer, and they were then hung on a clothesline on the southwest side of the house. Washing clothes was a major job in housekeeping and one that seemed to be getting more and more time-consuming. We had eight of us, and even being careful and frugal there would be a pile of clothes twice a week or more. We decided that maybe we could get a gas driven washer to at least eliminate the hand-driven washer part of the operation and also the ringer. Asking around we found out that at Baytree, across the road was an old bachelor, and his cats, who had a gas-driven washing machine he was not using.

We went to the house and knocked on the door. The bachelor was home, so he let us in. I can see it all clearly; the house was dingy and dark and piled with all sorts of furniture and boxes and cat boxes, and it was so crowded that one could only get from one room to the next by a single pathway. He must have had 10 cats, and it was like getting hit in the forehead with a board, unexpected shock. How could the man be living here, and it was even beyond acceptable conditions for a cat. The odors were also stifling and depressing. It was dirty beyond dirt. He did show us the washing machine which he said was in good shape and hadn't been used for five years or so, maybe the last time he washed clothes or took a bath, eh? We really wanted to just leave, but he told us we could have the machine for $50. And it was just what we were looking for, a tub type machine driven by a gas motor, and it had a ringer and, below the dirt, seemed to be serviceable. We bought it, took it home, and scrubbed it out until we could see that there was a reasonably clear enamel finish on the outside and the inside seemed welcoming enough for our clothes. We would use it outside or by the window next to the pot belly stove, attaching an exhaust pipe on the engine to spew the exhaust gases outside. It worked out OK and was a labor saver for the time being. The clothes still had to be hung out to dry, although in very inclement weather, we had a rack on which we hung clothes inside. That occupied much of the living room but was a natural humidifier.

CHAPTER FIFTEEN
The Church and its Members

We immigrated to Canada based on our dreams of a better life, and our move to the Peace River Country was directly influenced by the people of the Baytree Mennonite Church. As I have written, Carol Peachey wrote glowing and intriguing letters about this country in the Budget. Carol was the oldest daughter of Jesse and Faye Peachey. They came to the Peace River Country some five years before. They were originally from eastern Pennsylvania, raised as Amish. They moved west, finally arriving in the Peace. They had six children when we were there: Carol, David, John, Twila, Julia and Glen. They seemed to be good farmers, raising grain and some beef cattle and doing other odd jobs. Jess had a portable sawmill that he used often during the winter. And he was a good backhoe operator. He was immensely helpful to us in digging trenches for our house foundation, for waterlines, for drain lines, and the lagoon. They had a comfortable house, a big barn, and a couple of other outbuildings. The Baytree school was at the west end of their property. They had a dugout for their water supply, and they told us it was not big enough. In late March and April, after a long winter of snow and ice-covered water, the water was brackish and colored nearly black. They needed to boil it to make it presentable. And there were squigglers in the water as well. But they did have electricity and a telephone, so they were in the 20th century. They were no longer Amish, since they had most of the worldly contrivances. They had a large van, maybe a 10 seater or so, a pickup truck, and a grain truck. And they also had two or three medium-sized tractors with an assortment of farm implements. They were the biggest part of the original settlement of the Mennonites in the Baytree area, and they were the leaders. Jess was a preacher and was obligated to give a sermon nearly every Sunday. The Mennonites were good singers and had Wednesday night hymnal services, mostly singing favorite hymns from the Mennonite hymnal book. And they dressed plainly in the sense that their clothing had no bright clashing colors. They were not averse to wearing bright colors, blues and greens, and even red occasionally, but it was clothing that covered nearly the whole body. The women wore a small head cover when they were in the public. and many of the women's dresses included a kind of apron over the basic dress. The men did not have beards or the cropped type of haircut. In appearance they could easily pass for regular chaps in Dawson Creek. Without the help of the Peacheys, we would not have made it through the first winter as well as we did. They were not loud, overbearing or demanding. They were good neighbors during our stay in Canada.

Carol Peachey was the regular schoolteacher for our children, and she seemed to be competent. She had a wide interest in biology, nature, history and the arts, and was able to also do basic arithmetic for the students. Physics and chemistry and algebra were a task for her, but some of the children learned some of these on their own. Being a parochial school, church-based, one always needed to be sure no one was offended by some of the more "controversial" subjects like evolution, the reformation and history, astronomy, biological processes, and technology. Surprisingly enough, the Mennonites embraced technology for their farming and vocations. Television was officially out, and so, significantly for us, were musical instruments. Why David (the king) could play the cymbals and dance naked in a parade in front of his troops yet we were not supposed to play a piano became a point of trouble. Even listening to some of the contemporary music was not acceptable. I never pressed it at all, but I wondered if Beethoven, Mozart and Hummel were acceptable, probably not. But Carol was a good teacher, and our children listened to her and respected her authority. She never overstepped her bounds regarding discipline, which can become another point of contention.

The Baytree Mennonite Church was founded by the Peacheys and the Stutzmans, Willard and Arla. They were the older couple, having come up from Oregon, and Willard was the official deacon of the church. He also gave sermons and was mostly the anchor for the theology and church member behavior. He could admonish members if need be, but he was a mild, quiet person and kind. Arla was the gardener of the region around Baytree. Well, she knew gardening all her life, I suppose, and they were the only ones who tried the grow tomatoes and corn in the Peace River Country. They succeeded, kind of... The tomatoes were red and hinted of those lush fruit down south but didn't quite make it. Their corn was marginal although it developed ears and was edible. Their farm was south of Highway 49, directly south of the Peachey homestead, and halfway up the Saddle Hills. They were on a northwest corner of two gravel roads intersecting, and I estimate that the elevation at their home was 200 meters above our farm and probably 150 meters above the Peacheys'. This was significant because whenever we had exceptionally cold temperatures, say -50 degrees F, Willard and Arla were -35 to -40 degrees F. Heat rises, and it seemed to almost always occur that way. It is of course not absolute but expected. Both Willard and Arla were quiet and soft spoken, easy to get along with and helpful.

The Sadlier family seemed to have moved up to the Peace three or four years before us. The father, Phil Sadlier, was retired from Montana, around Kalispell, and was drawing Social Security from the states. His wife, Vivian, was not healthy, and they had to cope with some major issues. But both Phil and Vivian were friendly, outgoing and helpful. They were raised as Catholics, so they had a different perspective of the Mennonite Church. They seemed mostly to accept all the theology and doctrines that they were asked to swallow. And they bought a quarter of a section just east of our first house, the Tomzak place, on which they proceeded to build a home. It was a typical one-floor ranch-type gabled house with a shallow roof pitch. On two succeeding days while we were still at the Tomzak place I helped Phil roof his with tar paper shingles. Well, they call them asphalt shingles, but it is nothing more or less than tar paper, good for 15 years tops. I suspect that roof has been recovered or whatever at least twice in the last 30 years. But Phil and Vivian were good neighbors as well, if a bit far away after we moved to Pilsworth Corner.

Jim Sadlier was Phil's son, and he and his wife, Kathy, moved up from Montana when Phil came. They were a big family, and I remember them having more than 12 children and with all the noises and helter skelter activities associated with that many young people.

I remember them trying to accept all the theology and doctrine and behavior patterns expected of a Mennonite, although I think they may have had some minor excursions into other areas. I found them to be curious about our coming up, asking us why we came, and I sensed some confusion on their part. But they were helpful to us and offered to keep in one their freezers bear meat that we had butchered. That was a big help to us as having some 76 kilograms of meat, even wrapped, outside would have been prey for wild animals, our dogs, or spoilage during warm weather. Jim and Kathy's farm was about midway been us and Baytree, on the old Highway 49 which there ran parallel to the new grade, about 400 meters or so north. They were on a corner with the Bonanza road. The Bonanza road was one of the few paved roads in Alberta at the time. It was paved starting at the intersection with the new Highway 49 grade and all the way north, some 12 kilometers to a small hamlet of Bonanza. I never could understand why that road was paved, but it was. Jim and Kathy would, during the following year, donate a corner lot to the church and build a new structure for the school and church. The old school was not quite good enough, and this was another shade of the past.

Paul Sadlier was Jim's younger brother, and he and his wife, Becky, also moved up with Dad. They lived east yet of Jim and north about six kilometers up a side road that stopped at their property. That was necessary since a large creek meandered through their property, obstructing any direct roads that may be planned straight north. The creek probably began in the muskeg that was at the far west end of our property. Someone said that the year before a "Cat," caterpillar bulldozer, was going across that muskeg during the late winter and just disappeared, cat and driver, never to be seen again. But that was just a rumor, I think. Paul and Becky had one child and would then have a second while we were there. Becky was raised in the McBride, British Columbia, area so was kind of at home around Baytree.

Both Jim and Paul Sadlier farmed some, raising beef cattle and some grain, oats and barley. And during the winter months they were in the bush, logging. Jim claimed to be a feller, a man who actually cut trees down, and Paul drove a skidder. I remember Jim expounding one Sunday after church services how he would go into a grove of trees, mostly spruce I assume, using a chainsaw, and make a cut about three quarters through the butt of a tree at the edge of the grove that he was attacking. He would then move to an adjacent tree and do the same thing. From there he said he would kind of fan out always going down from the teetering trees, and after maybe 30 trees had been severed three-quarters through, the last one would topple over, felling all the previous ones, like dominoes. The prudent, wise timber man would never do this, I thought. A tree would be notched about halfway through on one side and then cut through on the opposite side, always making sure that one was clear of falling branches and far enough away from the tree butt to avoid any possible jumping of that butt as the tree hit the ground. This method would be a bit slower than Jim's described method but safer. And working in the woods is dangerous enough even with all the logical procedures for safety. I never fully believed Jim's story but never told him that.

Paul, I remember, purchased a new red Massey Ferguson combine to do his grain. It must have cost him quite a bit of money, and I think he borrowed to get it, but that is another story. He farmed the quarter section directly north of our farm that first summer,

John Trinier was a church member who came to the group while we were there. John actually was raised in eastern Canada, and his affiliation with the Mennonites was marital; his wife, Irene, was a Mennonite whose parents were living yet in the McBride, British Columbia, area, Joni and Lydia Mae Byler. John was interested in draft horses and talked about horses with me often after church. He mostly seemed to make his living by driving a dump truck. It was an up-to-date, three-axle rig that could haul quite a load of gravel, sand or grain (if cleaned out). His work was periodic and uneven, like much of the work at that time. Even timbering was susceptible to unfriendly weather, too warm being the worse, as the forest floors would then be soft and the muskegs dangerous.

So early on the church was comprised of the Peacheys, Willard and Arla Stuzmann, Phil and Vivian Sadlier, the Jim Sadliers and the Paul Sadliers, John and Irene Trinier, and us. We were a significant portion of the active participants, and our four older children were a major part of the school enrollment. But we were not official members of the church, and that would never happen.

In late April, a family moved in from the Fort Vermillion area, the Schlamps. Ben Schlamp was raised as a Holdermen Mennonite in southern Saskatchewan and had somehow moved north and west, ending up near Fort Vermillion. This was a settlement about 500 kilometers north of Baytree in the broad Peace River valley as the river approached a conjunction with the Slave River before emptying into the Great Slave Lake. It was a good agricultural region, and I think Ben had moved there for farming. But somehow he then decided that he was a better woodsman--timber and sawmilling. So he sold out along the upper Peace River and moved to Baytree. He and his wife, Susan, had 14 children, but the oldest 10 were grown up to the point where they had moved away. So they had four children when they moved to Baytree in the spring of 1979. They had a mobile home, and they rented a parcel of land about two kilometers south of the Baytree store on a dirt road. They settled the mobile home on the land, close to the road after clearing out some brush with a bulldozer that they owned. Quickly they fabricated a large platform against the mobile home and built on that a framed building. I remember them heating the mobile home and the new addition with wood, a small pot belly type in the new addition and an air-tight stove in the mobile home. Ben always seemed to be a man on a mission, and right then his mission was to get set up so he could scout around for some timber to harvest and set up a sawmill to sell lumber. He was a gruff, heavy-set middle-aged man who looked just like a typical native of the country. He did profess to be a Mennonite, and I wondered. I thought he brought a fresh breeze to the church congregation as he was inclined to say things that I might have said, if I were a member. Since I was not a member, I felt as a guest and tried to behave that way. Ben was a bonafide member of the Mennonite Church so he could exercise all the privileges going along with that: taking of communion, reading Bible verses during the church services, and entering in on the various discussions of the church and the school. I think he would, on occasion, drink a beer or some wine, smoke a cigar or cigarette, and I even heard him use a bit of profanity. We visited with his family a number of times, and our discussions often turned to politics, the world situations, history of the world, and his take on the church. I think he was always ready to relax some of the tenets of the group; Why not play musical instruments, why not see a movie or even watch some television, and what was wrong with a bit of wild behavior? I had pretty much hunkered down into a work routine since we left the states, and Ben reminded me that life is not just following along with a particular group, particularly a religious group professing to be in the right know. I was always, even when Joy and I were Lutheran, Church of the Brethren, or Old Order German Baptists, rather lukewarm. My membership in religious groups was always at arm's length or beyond. The Lutherans I tolerated and was reasonably active as a Sunday School teacher for a while in the First Lutheran Church in downtown Dayton, Ohio, but I have always had some suspicions that Christ was not a member of any of these groups and would not be. I felt some vibrations that Ben was of the same persuasion.

Susan was a mild, quiet good person who was raised to obey her husband, and I could even then see that that could be very challenging for her. While I did feel some commonality with Ben, I wondered about him when he overstepped his bounds and antagonized some of the church members and when he was rough and callous with his children and wife. It was to be a constant source of friction in the church and in his family too.

Shortly thereafter, in early May, a young couple, Gary and Mary Davis, who were college graduates with degrees in psychology, moved to the area and stayed at the Peacheys' farm. They were provided with a small one-room house to the east of the farm buildings and along the dirt road. The Peacheys let them plant a garden and live free of rent, I think. They did not stay too long as they seemed to be just searching and wandering. They visited with us a few times, and we had some interesting discussions, much about their field of psychology. I remember them saying that the subject matter was very interesting and pertinent to people but not fruitful or able to provide any answers. And that was why they entered the community of the Mennonites, hoping to find some solid, satisfying answers to something.... I always wondered about this, and while I never studied any psychology in a formal organized manner, I did read Freud, Jung and even Nietzsche. It seemed to me that the writings were vague and often irrelevant. I did like to read Nietzsche, and I feel that his thesis that God is Dead is an ugly, truthful insight. And my reading of Nietzsche leaves me thinking that he is not the anti-Christ or whatever evil some would think but is just stating that those who profess to worship God and Christ are actually killing him. They profess to believe, but their actions speak otherwise, and they make a liar of Christ. I am not sure where Gary and Mary went, but one day they were gone.

As spring was turning into summer, Jim and Rachel Hackman arrived from northern Wisconsin. Rachel's father was a preacher in a Mennonite community in Wisconsin, and somehow Jim and Rachel either knew the Peacheys or, like us, were intrigued by Carol Peachey's letters and decided to give it a go. They had two little children at the time they arrived, and so they set up housekeeping in the Peace River Country. They had some financial resources which they used to purchase a quarter section of land, and they quickly put up a house and a couple of outbuildings. Jim was planning to farm and maybe do some logging with the Sadliers or Jess Peachey. As I gathered, Jim was not raised as a Mennonite like his wife Rachel, but was willing to do it. They both were young, not more than 30 years old, and healthy. They were pleasant and seemed to want to fit in.

And then came David Donaldson and his wife, Annie, and young son, Hilding. They too were seeking something, as we were. They came from east around Wanham or even farther east, and they came driving an old yellow Toyota four-door sedan. They settled into the quarters vacated by Gary and Mary, and they were to be around for the duration of our time in Canada. David wanted to be an Amish. Actually he described how he wanted to use horses instead of gasoline vehicles, and the Amish seemed to be the tower of Hope or maybe Babel for those seeking a simpler, meeker, more sensual life. David had education, although I never knew if he attended a college or even if he finished secondary school. But he did profess to be a teacher, and he volunteered to help at the school.... That would come.

So during the spring and early summer of 1979 a number of people appeared and participated in the Baytree Mennonite Church and school. I think this may have been the spur for planning and executing the construction of a new church and school. And it was coincidental, or was it, that all this happened during our first few months in the community. We had planned this for three years and more, but what about the Schlamps, the Davises, the Hackmans, and the Donaldsons?? It was becoming an interesting community.

And during the late summer Joni and Lydia Mae Byler moved up to Baytree. They were the parents of John Trinier's wife, Irene, and they had lived in McBride, British Columbia, for some time. But they were being encroached upon in McBride by all sorts of people. Between the timber and lumbering industry and the oil rigs beginning to move in, plus always prospectors for various minerals, including coal, it was just getting to be a rough, unpredictable community, rather like a frontier town. And Baytree was still far away and quiet and not experiencing the prospector types and the roughneck oil rig people. And they were part of the overreaching Mennonite community. Joni was a preacher and actually was the bishop in McBride, but he and his wife moved to Baytree. They purchased a half section of land some 15 kilometers north and five kilometers east of our place. They were on a knoll looking west toward British Columbia, and they had a mobile home at first. They planned on building a substantial home and intended to farm, raise beef cattle and grain. So by the coming of the winter of 1979-80 the church was bulging with members, prospective members, and participants.

CHAPTER SIXTEEN
Neighbors

Not to be overwhelmed by the encroaching Mennonite community, the settlement here at Baytree and the Peace River Country had a rather extended and interesting history. The area was seriously settled in the middle 1800s by eastern Canadians, mostly English and Scots. In the early years of the 20th century a significant influx of immigrants from eastern Europe, Poland, Czchoslovakia, the Ukraine and Russia came to partake of the rich land. I am sure many of them thought this was like home without the Czar, the King of Poland, the Germans, or someone giving them trouble. The country is at a latitude comparable to Siberia or the Russian Steppes. And the weather was also comparable, I think. After the First World War some Germans came, and the eastern European influx continued. Then before and after the Second World War there were more of the groups coming. In the United States it is assumed that everyone wants to come to the USA, the land of the free and the home of the brave. Many people do not consider the US to be a land of brave or free souls, and Canada has a strong appeal for having free and brave people. Reading between the lines and talking to people who have experienced both Canada and the US, it should not be a surprise that many people consider Canada to be the more desirable. So the natives, the plain people, or if you will, the non-Mennonites, were an interesting and friendly community. It was a diverse group, and along with all these nationalities, the languages were not stamped out as was expected in the US. In the Dawson Creek newspaper stores, or even in the groceries, foreign newspapers were available--German mostly, but also some Polish and Ukrainian publications. We were to make friends with a number of the natives or people who were here before the Mennonites. Like all peoples they brought all the pros and cons along, which we were to find out.

Directly across the road from us, on the south side of Highway 49 lived Fred and Irene Pilsworth. The corner right there was called Pilsworth Corner, and it was probably not named after Fred and Irene. They had a telephone which we could use at any time, they had chickens and eggs which they sold to us, and they were always willing to sit down and talk about the area--its history, its people and the weather. Their house was unpretentious but comfortable. I remember it being maybe four or five rooms with the center being a family, dining area. Fred and Irene were both probably in their 70s when we were there, and they seemed to be getting by. They were active, always having a nice big brood of chickens in a coop to the west of the house. Irene seemed to be the chicken caretaker, and she demonstrated by my extended observations that the key to keeping the chickens warm was to have just the right number of hens in the house. If there were not enough, the hens would get cold and not lay eggs, and if there were too many they would get hot and again not lay eggs. Irene was a native and knew the region so well that she could gauge just the right amount of chickens to keep in the house. Of course she was not averse to thinning the flock down by butchering a hen for a Sunday meal if there seemed to be too many.

Fred was mostly active with cribbage. He was often sitting at the dining table with a neighbor, Reggie or Dorene Higgs, his brother Albert, or just a passing neighbor playing a game of crib. Fred would wander outside when necessary but mostly was in the house and visiting and giving advice. I did spend some time talking with him as he seemed to be a good source of local information and maybe even for some contacts for work, sales or whatever. I found out that he had actually lived in the log house we purchased. He was raised there and farmed there for quite some time. He described working with horses and that once a month he would make the trip to Dawson Creek for supplies with a wagon, or a bobsled in the winter, and a team of horses. It was a good 45 or 50 kilometers to town, and that would take around five hours. That would be 10 hours or more of travel, so he would stay overnight in the Alaska Hotel or another of the local hotels. The next day he would head back home. And that was only maybe 15 or 20 years before we came, so it seemed that the region and its people and the behavior patterns were changing rapidly. He admired my horses and came over often to look at them and comment about them. Fred seemed to be a good horseman, although he never volunteered to physically help.

Across the road from us, to the east on the corner was Albert Pilsworth. Albert was Fred's younger brother, and he had a one-room house and a set of granaries. He seemed to have been a grain farmer totally, no chickens, cattle or horses. I remember that he had the typical set of farm machinery: two moderate-sized tractors, three or four hopper wagons to haul grain, a disk harrow set, and a combine. I don't recall him having anything else, but he probably did. We did not get to know him very well as that first winter he left for the Okanagan in January. The Okanagan Valley was kind of the banana belt for Alberta and British Columbia. They didn't have bananas, but the valley was right on the southern border of British Columbia along the Okanagan River sandwiched between the Rocky Mountains on the east and the Coast Range on the west. Excellent fruit, peaches, apples, pears, berries of all types, and vegetables were grown, and the winters were moderate with maybe once a month a fleck of snow but normally above freezing and lots of sunny days too. Albert must have been doing this for some time as he was single and had no good reason to endure the Baytree winters. He would return in the spring, after the ice break up. And during the next summer, 1979, he would sell his half section to a younger couple, Roland Gauche and his wife.

Up the road, about a mile or so north of us on Twelve Mile Road, were George and Irene Blakus. As I have written, they were the ones who sold us the quarter section at Pilsworth Corner, and they visited with us often. George had a blue Chevrolet pickup truck, and it had a thermometer attached outside the cab by the rearview mirror. He would drive up our lane and see how we were, talk a bit, often sitting in his cab if he was in a hurry, and we could get the most up-to-date weather temperature, 30 below or whatever. But then again, he was always able to spend an hour or so sitting at a cup of coffee in our kitchen and discussing the recent news events or the farming conditions. I remember his house sitting in a clearing but up against a woods, or bush as they called most of the forest there, to the north, and they had a few windows looking out to the south where there was cleared land. The natives all seemed to like to have an open horizon somewhere. We became good friends.

And one of the early neighbors, from renting at the Tomzak place, were Nick and Rose Kanda. They were probably in their late 40s when we moved there, and they were natives of the region. They knew all the weather patterns, the quirks, the forests, and the land itself. Nick told me one day in a quite casual way that during both the world wars, the Peace River natives were not the most patriotic, and there were many who just refused to enlist or show up at the recruiting offices to go to war. Why go someplace and meddle in other people's businesses or problems when they hadn't done anything to you? Nick had a brother who lived on a section of land to the south of the Tomzak place. We never met him but often saw him out working in his fields. Nick and Rose were folks who just made us feel at home, without any theology or liturgical lines. Joy became a very good friend to Rose, and we received lots of good advice from them both. They had at least two daughters, one who was still at home and about Heidi's age, Kathy. Kathy and Heidi became friends. I remember Nick selling me a few buckets of barley now and then to feed to the chickens or the pigs when we purchased some.

To the east of our log home and just east of Pilsworth Corner was the half section of Werner Kwazoski. Werner was an older bachelor, and he kind of befriended us. He drove a well used Volkswagen bug and would bounce onto our property maybe about once every week. I think he was lonely, and he would visit with us and, when he stopped after making an errand to Dawson Creek, would have some candy or cookies for the children. We learned that he was a war veteran from the Wehrmacht, the World War Two German army. I never pursued my interest in how he managed to survive that awful event. My reading of that blot on history is that Werner must have been on the western front, probably fighting against the English and Americans and Canadians. If he had been on the eastern front, in Russia, the Crimean peninsula, or the Ukraine, he would have surely perished. His surname was Polish, and I often thought he must have been an East Prussian, with maybe a Polish father or grandfather. It is so tragic and I believe stupid that people have no ideas of history. It is always assumed that our side (whatever that means) is right and the other side is always wrong. The news and stories coming from the years of the Second World War just make it so simpleminded, if you will. And even before then, I feel that most people just want to get along, and the Germans and the Poles were living side by side whether in Germany or Poland, and except for the politicians and greedy business people it must have been somewhat peaceful. Of course many people make a big deal that our society is a Judeo-Christian culture with Greek and Roman influences. I have a hard time respecting all of them. The Greeks and the Romans and the Israelites, and the Christians too, were so militaristic; they originated, or at least spread the big lie, the eighttth deadly sin, about "laying your life down for your country is the best thing you can do." Werner was just a kind, quiet person. I know that once a month he received a check from the West German government for his service in the army. His house was maybe a half a kilometer east of the intersection at Pilsworth Corner and was one of the first authentic envelope homes I was in. An envelope house is a house inside a house, and actually Werner's was more of an accident or just a prudent arrangement. He had a cozy living quarter right in the middle, surrounded by other rooms that were not occupied during the winter cold. So he heated only the one or two rooms of the center, and the peripheral rooms were just there, bleak, dark, dusty, and storage space. But even though they were unheated, the temperatures moderated from the outside so it was a natural insulation of the living quarters. Being a bachelor, he had plenty of space and his own agendas. The house was close to Highway 49, maybe 50 meters off the road, and surrounded by some volunteer poplars and a few birches. He seemed to be of the type who preferred having trees around the house, and it showed. He had a couple of old granaries, but his land was rented out to neighbors to grain farm. Werner was just there. I went to see him a few times. Joy, I think, thought kindly of Werner, and when she baked some cookies or a cake, she would wrap some up for Werner and I would take them to him. He was appreciative of it but only in his quiet subdued way. I wondered if he thought we had some agenda for him. But here I am probably straying too far afield. So Werner was a good neighbor, nearby.

George Fitzpatrick lived up Twelve Mile Road, the second half section north of us on the same side of the road. He had a family, wife and children, and just kind of came to me one day asking if I would be interested in baling his hay. He had maybe a half section of land, and I really did not get to know him, other than from his lush hayfield. But I think he was not a full-fledged farmer but had a job of some sort in Dawson Creek He was quiet and not one to expound on his accomplishments, plans or whatever. I liked him.

Reggie and Dorene Higgs lived in a mobile home on Fred and Irene Pilsworth's property. It was kind of tucked back off the road. And actually Pilsworth Corner was a crossroad. Highway 49 was the east-west road, and north and south was Twelve Mile Road. Officially Twelve Mile Road ended at Highway 49, but a dirt one-lane road went straight south up the hills toward the Saddle Hills. It actually went maybe five or six kilometers south to some oil rig clearings that had not yet been developed. About two kilometers south of Highway 49 was the evidence of a grade, an old railroad grade running east and west made during 1914 or so. It was to be the main railroad to Dawson Creek for the Alberta Northern Railroad, but when it seemed that steel was needed for war supplies, the railroad did not lay any track west from Spirit River, and the grade has just been there ever since. The railroad instead went south from Rycroft, some 15 kilometers east of Spirit River to Grand Prairie, and then that line wandered west again to Dawson Creek. There were probably some neat political games played in all that maneuvering, not worth going into. So Spirit River was at the end of a railroad spur west of Rycroft.

Anyway, Reggie and Dorene were probably in their late 50s when we were there. Reggie drove an 18-wheeler, a tanker hauling oil or something. One could always tell when he had a driving job as his truck, a diesel, would grind away for a few hours before he left. Sometimes he parked the rig right on Highway 49, but more often he parked it on the dirt road just south of 49. And Reggie would then be gone for sometimes a week or more for his kind of tour of duty.

Dorene was one who would visit with the Pilsworths, and she did do some small woodworking projects. She liked horses very much, and during the second winter she became a regular, coming over every day to help me harness the two horses and drive them up the road to get hay that I had stacked from the summer of 1979.

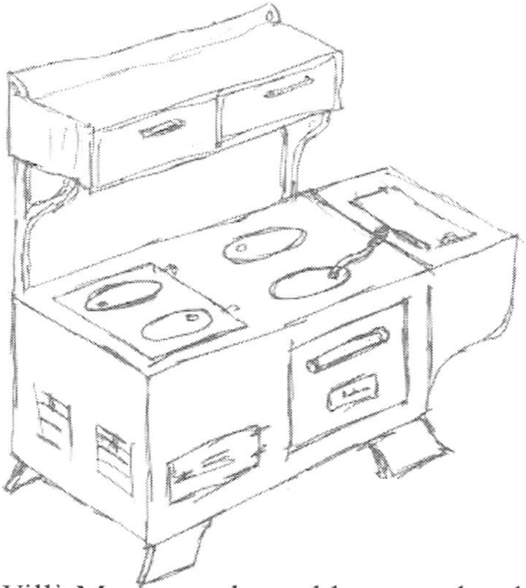

And Willi Meyers, who sold us our log home, would come over once in a while. He was between positions, easing out of the farming, and he would take over the Co-op petroleum station at Baytree during the summer of 1979. We got along very well with Willi although his family was always somewhere else.... Willi and his four-wheel-drive John Deere was the second tractor that pulled our house to our land on that big moving day.

Marvin Meyers lived just south of Willi in a double house trailer type of house. Marvin also farmed, and I think he may have done some other things also. He let me make some of his hay during the summer, and I appreciated that. It was done on shares, but was another bit of hay for us.

Roland Gauche and his family moved into the Albert Pildsworth place late in the summer of 1979. Roland was probably in his early 30s at the time, and it seemed that he was intent on continuing the farming operations that Albert had done. Roland's wife was sick and didn't make any outside appearances, but Roland came over to our house often. He was a bit outspoken as I remember in that he was a Jehovah's Witness or some sort of other group. He would discuss some of the theology of the Mennonites and Amish with us and always tried to get us to go to his church, which was in Dawson Creek. Kind of a soft-sell about why we were wrong. But I had no problem with Roland, and he was a neighbor who seemed to be trustworthy, as long as I didn't tempt him.

Ian Marchuk was a good friend who was about our age. He lived about four kilometers east of us on Highway 49. He had, I think, a half a section of land on the north side of the road and had a small two-room house, frame. A box-type stove was used to heat the house, and being a bachelor at the time, everything was simple or maybe disorganized. He had had a live-in girlfriend who moved out shortly before we moved to Canada, so he was indeed living by himself. He visited with us often, having no others to worry about and also to reduce the loneliness. Ian was from eastern Canada and had just migrated to Alberta in the same way that others had, including us. He seemed to have a bit of a conservative political outlook, if that is the right descriptor. He often talked a bit disparagingly of the governing Prime Minister of Canada, Pierre Trudeau who was a "liberal." As I have written before, I don't seem to know what the words Conservative or Liberal mean except I have a feeling that, as the double speakers of George Orwell's *1984*, a conservative was not a conservative, and a liberal was not a liberal, maybe the other way around. But we all thought highly of Ian as a good neighbor and that he was interesting. He seemed to have studied at a university in Quebec or maybe Ontario and was inclined to discuss not only politics but the economic situations and the world order and whatever else. He had two Belgian draft horses, and that alone made him a good friend. Anyone who had draft horses had to have something approaching intelligence. I remember him having a rough pickup truck with a cab, just a typical vehicle for the area. At that time the pickup trucks all were rear-wheel drive with the differential. Four-wheel drives and ATVs and all the paraphernalia of the early 21st century were not popular or even available. And most people thought, probably rightly, that a four-wheel-drive vehicle was an invitation for getting really stuck in mud or a snowdrift. Ian had Ukrainian ancestry and probably some other stuff, maybe English or Scottish or Irish. We saw much of Ian during the next two years, and it was always a good visit.

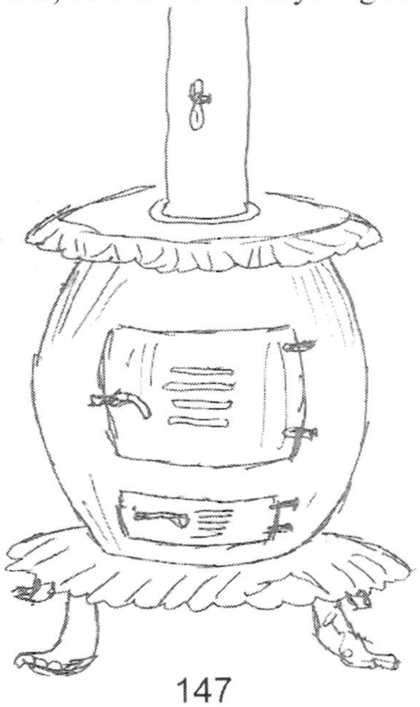

Arnold and Bunny Hayward lived west of Baytree about two kilometers along Highway 49. Arnold was the official snow removal engineer for the Province or town or whatever. They had a farm and also owned some land east of us on Highway 49, maybe three kilometers, kind of next to Ian Marchuk's place. We would make some of their hay on that parcel during the busy summer of 1979.

Gordon Jones was school teacher at Dawson Creek and lived near Gordondale, about 12 kilometers east of us. He had a good drive every day to Dawson Creek, and it was not until maybe the spring of 1980 that we really got to know Gordon and his wife. They were about our age and interesting and friendly people. Being teachers they did have some ideas on education and of course politics, social orders and the like.

Gary Henderson was, on first take, a local cowboy. We became acquainted with Gary during the late summer of 1978. He was indeed a cowboy, wearing a typical cowboy hat, even in the dead of winter, the denim pants, and a denim jacket. In the winter he wore heavy clothing like the rest of us. But he was a short, thin and wiry type. At first he didn't look at all able to ride a horse, let alone lasso and brand a calf. But he was a real cowboy from Montana, and he just seemed to know all the necessary moves to make. He even had that country dialect, not quite a Southern United States drawl but not the crisp Canadian banter either. We found out that he could play the guitar, and that just seemed like a given. I don't think Gary was a tobacco consumer, either smoking or chewing, but he might have been. He came one day because he found out we had two sons who could work and he wanted to build some fence above the Peace River. I think at that time Tim was not quite available as he was studying in school so Kurt went with Gary for a full day's work. They traveled north on the Rolla road, through Rolla and another 20 kilometers or so north to the Peace River, took the Clayhurst ferry across the river, and another 20 kilometers to Bear Mountain, where Gary's spread was located. Rolla was a small settlement, roughly six blocks by six blocks square with probably 30 homes and a general store. It was in British Columbia but only 10 kilometers or so west of Alberta. The Rolla road intersected with Highway 49 two kilometers east of Dawson Creek and was a straight shot north, all blacktopped. Kurt had a good experience with Gary and picked up some interesting and practical techniques for building fences. That is always something good to know, and I think Gary was impressed with Kurt as a person and as a helper. He would have him come to the farm often after that first time. Kurt was a good Bible and religious student, and he found out that Gary was a practicing preacher. He was trying to establish a church in Rolla at the time.

Tim also did some work with Gary building fences, sorting cattle, and typical work associated with cattle raising. As with Kurt, I know that Gary was impressed with the work ethics and the pleasant personality of Tim. All this of course was not my fault, maybe Joy's, but I really wonder how I was so lucky to have such good children, recognizable by others too! Gary and his family were to become acquainted with the Mennonites, through us I think, and they came to many of the church parties and special programs. They were good neighbors, even though they lived quite a way off. In Canada at that time distances sometimes were not a concern. Gasoline was not too expensive, and the pick of neighbors was thin. So anyone within 50 kilometers could become a good neighbor.

Adrian de Bruin was not 50 kilometers off. He lived near Willard and Arla Stuzmann and had some goats. We had raised goats in Ohio and were thinking of doing it again in Canada, so w e visited Adrian but did not go that route. We thought about it all and didn't want to get too thin too soon. We just had enough other activities and so did not do the goat thing.

Peter Kut and his family lived north and a bit west of us, maybe 20 kilometers or so. Peter was the man with the versatile four-wheel tractor that pulled our log home from Meyers' to our farm. Peter was really a fine person. He was quick to volunteer to help us, even though we were strangers and he must have known that we were kind of different. But he took us at face value and never was moody. He was friendly and cheerful any time, predictably nice.

Nelly and Bill Gamble lived north of us on Twelve Mile Road maybe five kilometers. Nelly was a Mexican, and I think Bill had met her in Mexico. I often wondered if Nelly really missed the warm, drier weather of Mexico. Canada, particularly in Baytree, was so much different, but she always seemed friendly. Joy was to be good friends with Nelly.

And there were others who we met and visited with. At that time visiting neighbors was a given. It meant going to someone's house, spending a few hours talking and having a cup of tea or coffee, some cake or cookies, and just getting to know the people. This was something that was mostly lost in the US, at least where we lived. Oh, some visiting was done but very cautiously and infrequent. I always sensed that here in the Peace River Country it was done because everyone was part of the community and everyone ought to included, meaning visiting them. During our time there, the community was changing fast, even faster than the locals could see. The oil rigs were moving in an aggressive way, and many outsiders, young and irresponsible people, included drug use, alcohol, and tobacco, and probably prostitution. That was very much not displayed in our community though.

But the neighbors helped us in physical ways and in spiritual ways too. We had a lot to do and also a lot to be thankful for. And thinking about the diversity of the region, some of the dialects were interesting. I mentioned that Gary Henderson had this distinct cowboy, western, country drawl, not the typical Canadian speech. The natives or home-grown folks, had some terms that really made them typical. *Eh,* or *A,* was a word that was included in almost every sentence or phrase. "How are you doing, *eh*?" "It sure is cold, *eh*?" And on and on. The truck had a *battery* and was made of *I-ron.* And I have mentioned already that the *Fullman* was once a lunar month. That may have been a transplanted term as not everyone saw a *Fullman*. The real Canadian, probably tracing their ancestors all the way back to England, pronounced *sch* as *sssch*, not *sk* like I had learned in school. So school was *Ssschuel, eh?* Schedule was *Ssschedual, eh?* And to scheme was to *scheeme, eh?* I suspect other parts of Canada had other equally interesting fallouts.

CHAPTER SEVENTEEN
Mellowing Horses

As the snow melted and the first spring in Canada became warmer, we had the feeling that we were just now really ramping up to what work was. The first eight months of our lives in Canada were challenging and interesting, but now we had to decide how we would try to survive and sustain ourselves in the community. We had a good warm house, a set of buildings, a good cow and two horses that were healthy, a small band of chickens, and some solid experience under us. So we needed to get to work. One morning in late May, when the trees were beginning to leaf out and the sun was hanging higher than usual in the sky, I heard the sound of what seemed to be a steam engine. Any person who did not live during or listen to the sounds of the years before 1960 would not understand the music of steam engines. Actually this was a thumping sound that started slow and accelerated to a rapid thump-thump-thump, before subsiding and returning to silence. I kept thinking of a steam engine being loaded for sawing timbers or maybe a thrashing machine separating a large bundle of grain and stalks. A railroad steam engine was not possible here. It would continue for some time, maybe 15 minutes or more. After a few days, it finally came to me; there were Roughed Grouse in the woods and the males were beating their wings against a log to attract the attention of the females. That was good to know, that we had Roughed Grouse in the woods nearby and that they had some ideas of multiplying.

On May 26 I finished a rough fence for the chickens and let them run around in the yard. They were content in their new house, and now they appreciated getting into the sun and outdoor air. But the next day it turned cold and cloudy, and in the afternoon it started to rain, a cold damp rain. By three o'clock it was snowing again, and the ground was white by 10 in the evening. Even with the long daylight hours now, the snow was the only light, and it silhouetted the ground, the trees and the dugout.

And in late May and early June there was a profound motion to everything. The frozen ground was beginning to thaw out, and that of itself was enough to create some new problems we had not yet faced. As I have written before, this country was underlaid with clay. On top, if one was careful, there might be as much as a half a meter of good loamy soil, decayed organic matter and fibrous material. It would grow good crops when the temperatures were above freezing. But below the loam was clay, good dirt as the highway people said. I don't think they knew what good dirt was, but the clay here did make for embankments, undergrades for roads and railroads, and even good building foundation support, as long as it was not wet. And that was the rub here. In the meltdowns of early spring there was a lot of water, particularly if there had been a good amount of snowfall during the winter. And we had plenty of snowfall during the winter of 1978-79. Earlier on, when we purchased our dear cow Flossie, there was a bit of a meltdown due to a Chinook, and our lane, exposed underclay (good dirt?), was treacherous and nearly impassable. And that was when only the top surface was greasy. Now, as the meltdown was in full swing, we had to learn driving all over again. The main Highway 49, if that is the right term for that clearing, had been prepared for paving during the last summer and fall. It was a good wide grade and lots of clearing on both sides, but the road surface was that good dirt, some scraped from our new dugout, maybe 6 or 7 meters below the woods' floor. It was the epitome of tenacious, viscous fluid. It was not solid like the frozen road surfaces that we drove over from October until middle May. That was easy to negotiate, even though I had had my doubts about traveling during the winter months in such far north land. It was because they did not have high winds to drift the snow around. As far as the frozen dirt roads, they were relatively smooth and firm. But now, with the meltdown, driving on the mud was something I had never before experienced. It became readily apparent why the use of four-wheel-drive vehicles was not considered seriously; if a two-wheel drive vehicle would get hung up in a mud bank, a four-wheel vehicle could disappear in the quagmire. So the key finally was the pickup, and Joy, Kurt and I could kind of get along on Highway 49, which was necessary if we were to get our children to school, get to town for some major supplies, or go the church. It was this: Drive slowly and don't turn the steering wheel sharply at all, anticipate at least 100 meters ahead what to expect and prepare by keeping the vehicle in a straight path, if possible. Highway 49 was wide enough to forgive transgressions of these basic ideas, but not much for not too long. One would find themselves on the side, down the shoulder embankment, or even sliding against some trees 100 meters from the road if the driver (?) had been driving too fast. Actually we had experienced all this in western Saskatchewan coming up when we were rescued by a road grader who dragged us through a wet quagmire over just a kilometer or so. And that was during the dry season after a rainstorm. These road conditions were somewhat like the icy roads, 1- or 2-centimeter-layered ice on the roads in the Midwest of Indiana and Ohio. Joy and I had grown up in that area and knew what the potentials were. But this was even more challenging because icy roads are just slick, and it meant sliding if the vehicle was not under control. Up here in the good dirt of Alberta it was slick as well, but the material was tenacious. It stuck to the wheels and globbed up to even keep the wheel from moving properly. Once turned sideways, a wheel could get stuck there, and any attempt at a corrective measure was doomed to failure. So we managed as best we could. We had a few minor cases of traveling down the middle of 49 and all at once the truck would veer sideways. It occurred for us when we were not going over 20 kmph, so stopping was not difficult. One had to display patience and going with the flow, so to speak.

And in early June Joy's parents, our children's grandma and grandpa, traveled up to see us. They lived in east central Indiana, and they flew, I think, out of Dayton, Ohio, to Chicago, to Winnipeg, to Edmonton. Joy's father and mother had never before been out of the states, so this was to be a first for them. I believe they had never flown in a commercial aircraft before. They rented a four-door sedan in Edmonton and drove the 800 kilometers from the airport in Edmonton to our farm. The major highways were all improved in Alberta (except Highway 49), so they had good solid pavement all the way. I think we advised that they go through Dawson Creek, as coming from the east would have meant they would have at least 50 kilometers of Highway 49 and its good dirt surface, all the way from Spirit River in. That could have been disaster for them. So they had clear driving, if long, from Edmonton to east of Dawson Creek. Highway 49 was paved in British Columbia up to the border with Alberta, and then came the dirt road. Somehow they managed the road from the border east to our place, a distance of some 50 kilometers. I think maybe that day the surface had dried a bit and made it not so tenacious, sticky or even viscous. We had a good visit with them, and it was good for us and them to see again some familiar faces. Bill, Grandpa, filled me in on how Chief was doing, and I could tell then that he was really enjoying the responsibility of taking care of a real horse. That pleased me as we knew he was having some serious health issues and his farming was going down. Taking care of a horse was just a nice hour or two during the day to pay attention to a sentient being, give them some grain, hay, and maybe a drink of water. I know he would then take a lead rope and walk Chief around some. The therapy of that surely helped him, and when we sold Chief later that fall of 1979, I wondered if we shouldn't have just given him to Bill. It may have extended his life longer. But we were able to keep them up in our house, and while it must have been a rude awakening for them, it surely brought back memories of their growing up and how things could be. They never admonished us for doing this (moving to Canada) and displayed interest in how another part of the world farmed and lived. They went to church with us on a Sunday, and we took them to the big city of Dawson Creek to view the obelisk at the center of town where the Alaska Highway began: mile zero of the Alaska Highway. One of those excursions turned rather bad. Grandpa was in his 60s at the time, a farmer all his life, and driving a motor vehicle for probably at least 50 years. And he was opinionated, stubborn, and displayed lots of self confidence, not so uncommon in people. He knew how to drive on Highway 49 and all the Alberta dirt roads; he had done it all his life. Sure... And he always insisted that he would drive his rental car. This one time we were headed west on Highway 49, I believe just east of the intersection with the Bonanza road. Nothing was out of the ordinary except that we probably had had a nice rain the night before or maybe that morning and the road was in an unforgiving mode. It was slick and gummy, and tenacious, and viscous all at the same time. He had learned not to drive too fast, that was good. And his long experiences in driving had given him the foresight to anticipate one vehicle ahead. Well, as quick as it could happen, the sedan, with all 10 of us crammed into it, just on its own diverged to the shoulder, heading almost perpendicular to the road. It all happened quick and quietly. The car did stop as it started to head down the embankment, and that was fortunate. Grandpa still felt he had everything under control and tried to back it out unto the middle of the highway, but it was all to no avail and actually making the car continue to turn around, much like a pirouette. I suggested that maybe I could get us going again, even though I knew my experience was not a whole lot more than his. After a few minutes he grudgingly let me take the wheel, and I did, luckily, slowly ease the car so that it first was heading in the same direction only 180 out of kilter, heading east instead of west. Ever so slowly we managed to get the car back in a straight line direction parallel to the road and on the good dirt, so to speak. Grandpa let Joy or me drive the car for some of the other excursions. I often wondered what he thought at the time about the roads up here. A few days later Grandpa and Grandma left for back home, and we wished them well. It had been good to see them.

The last week of May was more and more warming, and the snow was nearly all gone. As a fresh surprise, we began to notice large mosquitoes making their appearance. There were not so many, but they were large and could still be pestersome. I noticed one day that a puddle of water, fed from a sheltered snow bank that was melting, was full of mosquito larvae and that was probably one of their homes. So the large mosquitoes were the first wave of that part of society. By middle June the regular-sized mosquitoes would be plentiful and give us cause for swatting. And by August we would get the waves of "no-see-ems," multitudes of small mosquitoes and black gnats. They would be the worst. And one could swat continuously, and only in a breezing sunny area could you get some calm from those buggers.

And the first week of June the dirt was drying out and the roads were becoming more passable. I noticed that oats were sprouted and doing good as was the small patch of wheat. Surveying the cleared field, I thought now might be a good time to break Morning Star in on the unseeded part of our field. It was a bit risky in that Bess was in foal and would be expected to have it sometime in late June. Nine months and nine days is kind of the rule of thumb for horses, and she had been bred back in late August of 1978 to Conel Jay in Ohio. Nine months and nine days actually put it in the first week of June. It seemed that Bess was in good shape, and I thought she could get some good exercise without endangering the foal that was inside her. So on June 8 I took the green flatbed wagon to the field. I put the sideboards on it and put some 50 bales of damp, heavy hay in the wagon. That was to give it a good load and keep the horses from going berserk. I had been using Bess off and on since coming to Canada by herself with a set of steel shafts. This was fastened to the stub tongue of the green wagon and was really two steel poles extending forward from the tongue. Bess would be led between those two poles or shafts, and her harness would be attached. She had gotten accustomed to this set up and was reasonably comfortable. Everyone who had work horses, or any horses for that matter, told me that the more you handled and worked with horses the better they worked or drove and the healthier they were too. Horses, like people, were made to do things, not sit around or sleep. I had ordered a set of good new work harnesses for her and Morning Star some two years before from a harness shop in Holmes County, Ohio. I would now use them both for the first time and give use all three a workout. I wanted to be sure that everything was prepared correctly. I even put the harness on Morning Star a few days before and led her around for a short time to get her comfortable. And this all was coming none too soon. One of the major goals I had set for myself was to farm with horses. It all seemed so organic, so sustainable, and just so natural. I look at the "modern" farmers sitting in their cabbed tractors, probably listening to something they call music and churning along at 15 or 20 kmph making one or two passes over a field each year and calling that agriculture. It may be agri but not culture. Culture has to do with people, actually human beings, but there so few of those and billions of people that I suppose they too have some sort of culture. The tractor mentally and activity is agripornography with maybe a sprinkling of agribusiness. So real, sustainable farming and food growing will only come about when horses are part of the mix. I do need to be careful because too many people, farmers, in years past abused their horses, just like they are abusing their tractors now. But the horse, being a sentient being, can suffer. So we need farmers who have horses as their equal. That was what I thought of and hoped would happen with me.

On this day, June 8 I think, I fastened a two-horse shaft to the wagon. Kind of confusing, but the one horse shaft had two poles and the two horse shaft had one pole or tongue, between two horses; or the team. This single shaft, the wagon tongue, was nice strong steel, and at the wagon a double tree was attached to the tongue. The double tree was a set of two single trees, right, eh? A single tree was a short shaft, just a bit longer than the width of a horse and had stout hooks, one at each end. Sometimes these hooks had a clip to prevent a chain or another hook that would be attached to that hook from coming off. There were just a lot of clips and hooks and slides that had long histories of design and thought gone into them to make the whole harness safe, functionally, and finally practical. The single tree was attached at its center, lengthwise, to an evener by a pin or bolt or some other junction. The evener was actually the part that was attached to the tongue. So there are two single trees each attached to the evener, and the evener was then attached to the tongue. Here it is, roughly,

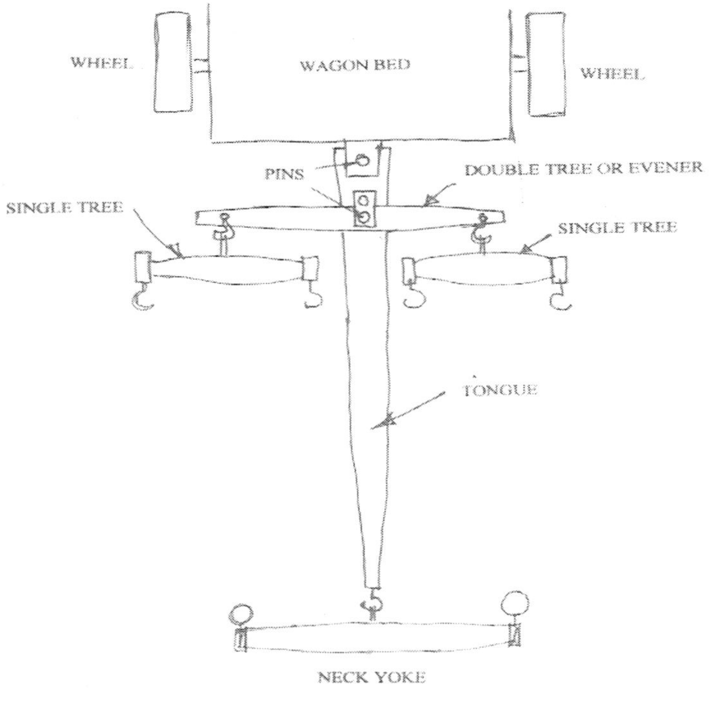

I had everything attached, the horses each had their harness on, and I led them out to the field for the run. I knew this was the crucial test, or drive, because if something went awry, the harness or shafts, or wagon broke, the least that would happen is that Morning Star, and even Bess who was a solid mare who knew about working, would be spooked for days and maybe forever, not able to work comfortably. So this was important, and I knew it.

I hitched Bess up first to the left side of the tongue. The hook up was final when a pair of traces, heavy leather belts, one on each side of the horse and fastened firmly to the harness at the front of the horse, was attached to the hooks in the single tree. Each horse had a heavy collar that was hung right around their neck and resting on their shoulders. The collar would be the real interface between horse and harness and wagon, and when the horse was harnessed up, the collar went on first. Then the harness with all of its component parts was slung over the horse's back. A set of hames, two wood or metal soft S-shaped poles, was part of the harness and these two hames were placed over the collar. Hames are often hung in fancy places as decoration, and even the farmers often put polished balls or something on their end at the top. They made the harness more flashy. The hames were actually the part on which the traces were attached. The traces were two heavy leather straps that extended from the coller/hames to the back of the horses, attaching to the single tree, one on each side. So the horse would pull against the collar, and this was transmitted to the hames, to the traces, to the single tree, to the evener, and then to the tongue and wagon. This provided the power to move a wagon, but there needed also to be the assurance that when the wagon and horses were going down a hill or the wagon had enough momentum to overrun the team, that was handled. Very cleverly I might say. At the front of the horse, at the bottom of the collar, a collar strap hung. It was attached to the collar through the harness hames, and this strap was attached to another strap, called the pole strap, that stretched under the front legs of the horse, split at the horse's belly area into two straps that were fastened to a broad, stout leather belt called the breeching seat or fold. The breeching extended across the back of the horse just about at the horse's hips. This breeching then would push against the horse's hips if the wagon tried to overrun the horses and prevent that from occurring. The front end of the pole strap was attached to the single tree at the front that was attached to the tongue. I have sketched the action going on in all this in the following figure, trying to give you an idea of how not only does one need to have a good sturdy means for pulling a wagon but also a well thought out braking mechanism for slowing a wagon down and not running a horse or two over.

A set of driving lines was attached to the bridle of the horse, one line on each side of the bridle. A metal bit was put in the horse's mouth that attached to the bridle as well, and this all was the communicating line between driver and horses: two long leather lines. It all was a good exercise in mechanics, the horse collars, hames, harnesses, traces and driving lines, and I often wonder how many engineers could hook up a team of horses, let alone describe how it all fit together and functioned. A good horse person knows all about this, yet I am sure would struggle and not be able to actually do any calculations to determine how much the horse or horses could pull, or how much stress is on the various components of the harness and hitching. I think back, and it all seems obvious now that there exists a wide, wide gap, a chasm, between practitioners, skilled craftsmen and farmers, and the designers and engineers. Accountants, business people and bureaucrats don't know anything but act like they know it all and take the largest shares of society's bounty. That has been going on now for thousands of years, I suspect, and will continue, until the meek inherit the earth.

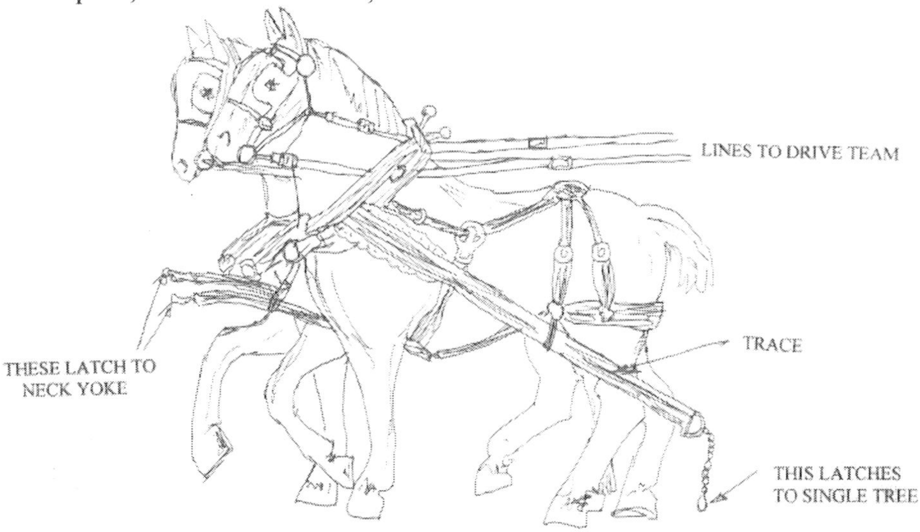

Bess was at the wagon, her traces hooked to the rear single tree, the front end of her pole strap fastened to the front single tree, and a set of driving lines draped over the wagon front end board. Morning Star also had her harness on, having been clothed in them a few times before. She had never, however, pulled a wagon or pulled anything behind her. This would be her indoctrination, the breaking, if you will. Actually I never felt that I was going to break a horse that must have weighed 10 times myself and had the strength of 20 of myselves. It would be a case of her getting accustomed to pulling inside a harness with a noisy wagon behind her. It was going to make her mellow, I hoped.

So we eased Morning Star against the wagon tongue and carefully hooked up the traces to her single tree and then fastened her pole strap to her side of the front single tree. I think we then attached the driving lines to Morning Star, and I was about ready to go. Kurt, Lori and Tim were helping at the time, providing extra hands and just trying to keep everything calm. I think I remember talking to the horses the whole time we were hitching them up and trying to act nonchalant, or at least calm and collected. Inside I was churning it up, anxious to go yet hesitant. So I climbed up in the wagon at the front end with a few bales of hay giving me some support behind. I thought that was a bit of comfort, and it never crossed my mind what would happened if the whole load slid forward against the front board, squashing me in the process. I would never have gotten this far if all the ifs were considered or recognized. We were on the north side of the field, the south side having the sprouting oats and wheat right at Twelve Mile Road. The children had each ahold of the two horses as I climbed up on the wagon, and at some sort of a signal, they let go. I gave Bess the go ahead, probably a sharp "Hai," and Bess started to pull. That dragged Morning Star along, and off we went. The reactions were so fast and furious that all I remember is holding fast to the two sets of lines, four in all, trying to keep them taut but letting them go. We had a full load, one which my tractors probably would have had a difficult time pulling at the front of the field, and we headed west toward the muskeg.

The muskeg was not quite a half mile, 650 meters, north of the road, and we just bounded toward that swamp. The ground had dried out, but here at the north end it started getting soft, and by the time we were near our west border, it was muck. The wagon went into the soft ground nearly up to the axles; if it had gone to the axles, we would not have gotten out. But the two girls just kept pulling, Bess in her seasoned and experienced way, and Morning Star probably just trying to get away from the terror of it all. I knew that we couldn't keep going west, and well before we reached the actual muskeg I pulled the left lines to get their heads turned left to turn the wagon. All this was just reactions on my part, but the wagon did gradually turn left, swinging though the swampy muck, and without even faltering or slowing down we turned south and then east, toward the road. It was going as good as could be expected, maybe better, because by the time we had made one complete circuit, the girls just responded to my command to again turn left and we swung around and headed west again. This time I had them turn around a bit before the muskeg, but it was still soft and I knew that any tractor, maybe even Peter Kut's Versatile, would have gotten bogged down, and the wagon would have been there for a month. But my good faithful horses did it, and after two rounds they were showing signs of being tired. I slowed them down, and we walked calmly around the front apart of the field where the ground was firm, and I could then tell, even though I claim no horsemanship, that Morning Star was ready for serious work. She was mellowed, not broke, as I would sometime later find out, but gentle. It was an experience like no other, and it trained me as well. I did notice that a pair of vehicles, probably both pickup trucks had stopped along the road, and they must have taken in the whole event. It didn't last more than 15 minutes, but what an adventure. I wonder what my children thought of it all. They told me much later that the Canadian adventure was the high point of their lives. That, I am sure, is stretching it, but that day, when two Belgian mares pulled a loaded hay wagon through a swamp, they must have realized that this was more like science fiction than fact.

The following week, on June 18, Bess foaled with a nice healthy stud colt. We called him Prince, and I wonder what he was thinking during our escapade in the field a week before. He had to have heard the rumble of the wagon, the tugging and puffing of his mother and sister, and then the other noises of me and the children.

CHAPTER EIGHTEEN
Making Hay in the summer

June of 1979 was warming up nicely now after a few false starts. And always one needed to remind oneself that this was high latitudes and summer was only one day removed from winter. One day in early June I walked out through the woods toward the oats field and was astounded to see that the sprouted oats were now 30 centimeters high and waving in a refreshing breeze. I made a mental note and realized that I needed to make sure my binder was in good repair, that the tongue would fit on the two-wheel dolly that was fitted to the binder, and that I needed to get some way to thrash the grain. I surely intended to use the horses to cut and bind the grain, but they couldn't pull a thrashing machine. I always could rely on good neighbors to thrash the grain for me, but that would be a compromise of my long-term plans. Actually they would be using a combine then, and that would circumvent binding, stooking and thrashing. And I also needed to get some equipment to make hay. I had taken a hay loader along with us, and it was available. The hay loader was designed to be hitched to the rear of a hay wagon, and it would automatically load hay on the wagon if one straddled a windrow of hay. We had used this loader in Ohio, pulling a wagon with my tractor, and it worked very well. It would load a hay wagon in short order. Then, with all that loose hay on a wagon, it was put in our barn. There we had a nice hay hook that would take a big bite of hay from the wagon, pick it up to the top of the barn where a track was installed inside the ridge. When it reached the track it would latch to a small four-wheel cart that ran on the track and shuttle the clump of hay either right or left in the hay mow. When it would reach directly above where the hay would be deposited, a long rope tied to the cart could be pulled from the barn floor and that would release the hay from the hook, and it would fall in a pile in the mow. We used Bess often to pull a long heavy rope that picked up the hay and moved it to the correct location in the barn. She was outside and would be led away from the barn with the rope in a single tree behind her. After the hay was dumped from the hook, we would either back her up to her starting place or remove the single tree and walk her back. It worked smoothly, and as long as the hay was only for our use, loose hay was preferable. And I had read in a horse magazine, or maybe an enlightened agricultural magazine, that loose hay was preferred over baled hay by animals. That was not what John Deere, Case, New Holland or the other farm implement manufacturers would want you to know, so it was discredited. And the college and university agricultural schools were essentially bought off by agribusiness, including organizations like the Farm Bureau, so loose hay was not ever considered. Besides, "it was too much work and cost too much," the standard line of everything that doesn't fit the mentality and plans of the powers that be. And they are there for sure.

Here in Canada I had no cavernous barn, in fact no barn for storing any sizeable amount of hay. I could store it in a pile in the field, making it rounded on top to shed water as best it could. But I think we decided for the short term to make baled hay and sell most of it. If we made enough hay to sell, like Jess Peachey and Jim Sadlier were doing, maybe that would give us a significant income to live on. As it was, we were beginning to see that living on our savings was not going to be sustainable, and this was a reasonable and quick way to get that turned around.

In the third week of June, actually on June 16, my father, Opa, came up to visit. He flew from Dayton to Chicago to Vancouver to Dawson Creek, which took the best apart of a day. He seemed to have enjoyed Vancouver, even though it was raining when we went through. Vancouver is at a spectacular geographical location on Puget Sound, on the Frazier River, and in sight of mountains all around. And it was a busy seaport; maybe it reminded him of his old hometown of Hamburg, Germany. Anyway, he stayed with us for a week and a half, and we had a good visit, although he never was encouraging about our venture. He could just not understand our rejection of all the modern contrivances of the age. Again, my opa, his father, would have fit in with us and understood our way. But my father was not of that persuasion and came to visit but not encourage. We made him feel at home and took him around to our friends, neighbors and even to church one Sunday. He was to leave on June 28, f lying back to Vancouver from Dawson Creek and ending in Dayton, Ohio. We took him to the airport rather early in the morning, and while we were in Dawson Creek we went to Kenver Equipment and purchased a used New Holland hay baler for $1,000. The baler was a model 68, an older model but one which would serve us well for our baling season. As with all used balers, it needed some tuning up, the plunger needed to be adjusted and aligned better, the cutting knives were dull and they needed sharpening, and there were some bolts needing tightening. But it was essentially a solid machine.

Two days later we purchased a four-bar International hay rake, side delivery type, from a Lloyd Smathers near Gordondale We paid

$250, and that was probably a bargain, for us anyway. That too would serve us well. I visualized it being useful even when we would go to loose hay and the hay loader. Windrows were very old technology. Actually I remember my Uncle Herman and Tante Mimi in Lemington, Wisconsin, having a dump type rake, and it also put hay in windrows back in the late 1940s. So this machine was doing the same thing, only a bit easier. You've heard that before too. But interesting enough, the older technology of handling hay was hand raking the cut and dry hay into piles and then combing maybe 10 or so of these smaller piles on around a pole in the hay field. If it was stacked properly around the pole, the hay would shed water over the wet seasons or snow during the winters. That is getting back to sustainable, but we are still a long way away from that; "too much work and too expensive???"

With the baler and rake I was now ready for some serious hay making. I planned on using the John Deere A to pull the baler and then haul the hay with the horses. And George Blakus had a field of hay that he thought was ready to make. In the Peace River Country one cutting of hay was about what could be expected for a summer. Maybe George's field could be cut again later in the summer again. But on July 1, Canada Day, it rained hard. The next day we cut his hay and on July 6 we had 627 square bales made and, of those, 600 were put under a black plastic tarp. We were a bit reluctant to do that, as with high humidity, moisture could get trapped inside and spoil the hay. We had made this hay on shares with George so he was entitled to half the hay. We kept the pile on his property, stacked rather orderly, and I would haul it, when time permitted, in the late fall or winter with the horses. I was planning on selling most of this hay, and as I write this all down and think about it, an important exercise that I didn't do well at all was marketing. Marketing was never one of my interests or goals. It should have been, but it was philosophically something that was distasteful to me. Why not let the customer decide what and when to purchase something? I always considered advertising as the lowest form of people's activities, and salesmen or saleswomen were the perpetrators of that evil. Well, here comes reality again, and if I had just advertised and made it known throughout the Peace River Country that I had lots of hay to sell, our venture may have turned out entirely different.

Meanwhile we were planting a garden in the field, and while it was quite a distance from the house, the ground was good, there was plenty of sunshine possibilities, and the ground looked like it would give us a good crop of vegetables. Joy, Lori and Heidi did much of the planting although we all pitched in.

Around July 14 we cut a good patch of hay along the side of Highway 49 and also along Twelve Mile Road, all near or on our farm. The weather was being cooperative, and the hay dried nicely so that on July 20 we had another 295 bales of good quality hay, mostly alsike and sweet clover. There was also some grass mixed in, and that would be plenty of hay for our horses and cows for the upcoming winter. From here on we could plan on using any more hay as a cash crop. And not to be too accommodating to us, the next day was a stormy one with a hail storm that killed 11 of our chickens that were nearly full grown. Such was the chain of events at our place. But we did go to Wanham that same day and picked up Bess and Morning Star. They had been about a week at Everett Parlee's being serviced by his stud Belgian, Bayou Bob. I didn't ask, but the name sounded like a southern US name, and maybe the horse had been imported from south. We now had both of our horses in foal, hopefully.

And again, in the last week of July, I went out to the oats field and was astounded at how high it was, over my head, and lush. The long daylight hours and the warm weather and the good ground just stretched the plants to their full potential. Now all we had to worry about was whether the grain would lodge. Lodging was the situation when the tall grain stalks would get pushed over in a high wind or even a pounding rain; it was then lodged and lying on the ground. But as I was going to bind and thrash the grain, even lodging would not be a loss. My binder was designed to efficiently pick up the lodged grain as well as standing grain and tie it in small bundles for shocking or stooking, as the term was applied here. With the modern combines and swathers, lodged grain was a challenge to harvest. A swather, as I was to find out, was a machine that had a wide front with a long cutting bar, a reel like the binders and combines, and a gathering canvas belt that put the mown grain or hay into a windrow behind the machine. The machine had a cab above the front so an operator could monitor the progress of the cutting, swathing, if you will. It had its own motor, so it was self-propelled. A neat tidy marketing ploy by agricultural equipment to have another machine to sell to gullible farmers. It did a tolerable job of windrowing grain or hay later on, but why? It was not inexpensive, maybe cheap, but not inexpensive. In the Peace River Country, thrashing was only discontinued within the past 20 years. Ironically, their combines were usually not combines but moving thrashers, as the grain was cut and put in a windrow by the swather and left to dry somehow that way. The combine, per se, did not have the typical cutter bar and reel but rather a pickup reel that theoretically picked up the windrow and fed it into the combine's separator chamber. But the pickup reel was not nearly as efficient as the old binders, so there was always a significant loss in harvest with lodging. As I observed it all in my iconoclastic or maybe curmudgeon way, here again was the agricultural machinery companies calling the shots. Cutting grain and laying it down on the ground to dry was not very smart compared to binding the grain and stooking it, or shocking it so the grain was standing up and able to actually dry out. And a grain binder was a rather simple, rugged machine with few parts to wear out. My binder was surely 85 years old and still operating dependably. But I am sure the businessmen could see that they could make a lot more money by selling a million dollar combine that would last maybe 15 years. And they surely must have seen a potential, upcoming market for grain drying as well; job security and maximizing profits are always the game with those people. I was not going to worry too much about grain lodging, and it turned out my binding and thrashing would be better than the more neo ways.

I have been using the term stooking to describe the placing of grain bundles in a vertical position to dry. This was an old way of letting cut grain dry very efficiently, and there are no better ways. One would take two bundles, one under each arm, and lean them up against the other one. Together they formed a steep inverted vee, V. Quickly then two more bundles were grabbed and placed against those first two, only 90 degrees from them. This was then the element of a solid shock or stook.

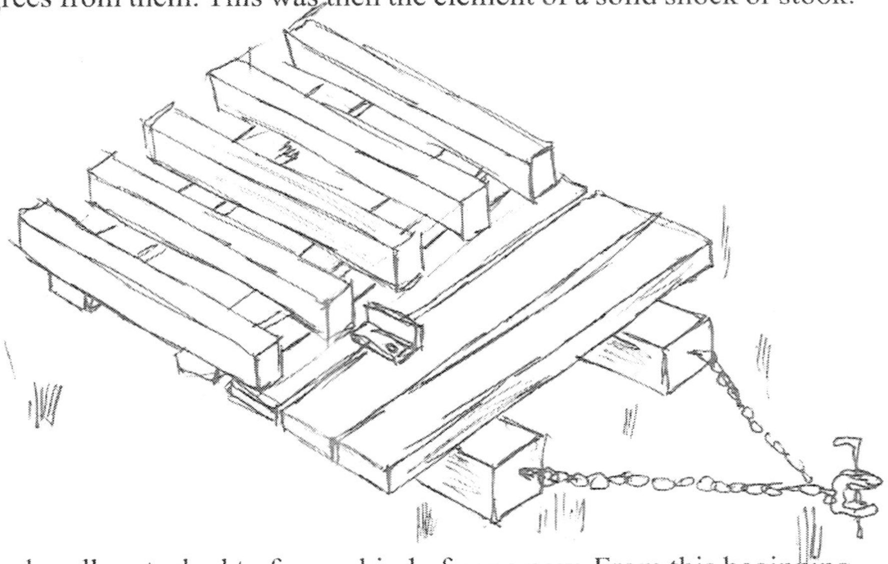

Four bundles stacked to form a kind of cone now. From this beginning, one could then place up to maybe 10 more bundles around the original cone to form a nice solid stook. Finally, it was topped with one or two bundles placed carefully right on top horizontally to form a cap and that would also help to shed water. A stook could last until the following spring or summer if need be, but I think the optimum was two or three weeks of reasonable dry weather followed by hauling the stooks to the thrasher for separating, or thrashing.

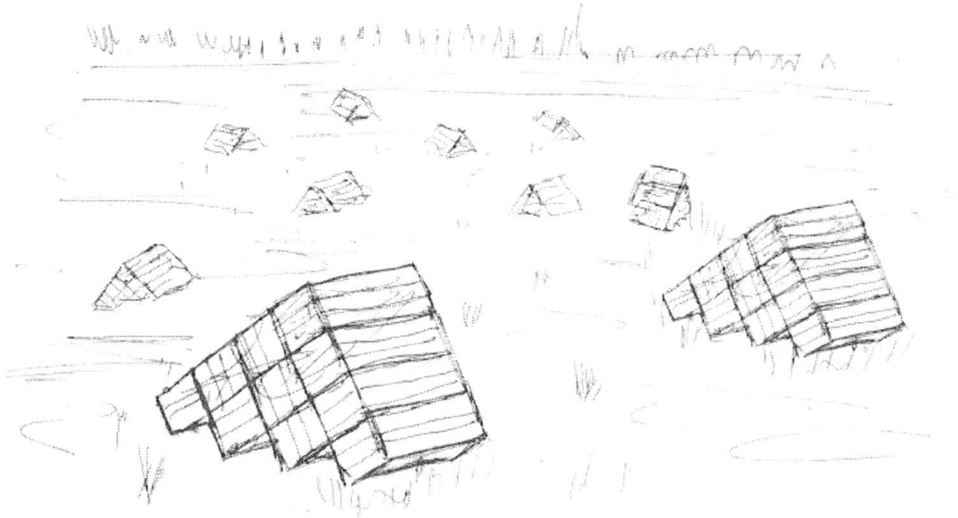

There was also another stook in the Peace River Country, the hay stook. While I put George Blakus's hay up in a large block pike covered by a tarp, one could also put 10 square bales of hay in a stook and leave this in the field much like the grain stooks. This was another ingenious method of just bending with the unique climate of this country and adapting to it. A hay stook was, as I said, 10 square bales of hay, four of them on a bottom row, placed at a 45 degree angle, tilted so to speak. Three bales were then placed on top of these four, fitted nicely between the four in the vee formed by the angles of the bales. Then two more were placed on top of the three, and one final bale, the tenth bale was put on top. The result was a clean, gabled stack or stook of hay. It would shed water as good or better than any other configuration, including putting a tarp over the hay, and the stooks could stay in the field all winter, if need be. Here is a sketch showing the simple but yet profound pattern of the hay stooks.

Now one may wonder how this configuration, stooking of the bales, could come about. That too was a simple yet profound process. One needed a stooker, and that would do it. The stooker was a sled with two runners on the ground or hay field. The sled was a narrow platform on the front, wide enough for a person to stand on it, and then there was another platform made of five 4-inch by 4-inch timbers, about as long as a square bale would be spaced at a distance apart so that a square bale could be placed between each of these short timbers. The five timbers were fastened underneath with a cross member, and about a third of the way back from the front, using the narrow platform as reference to the front, a hinge mechanism was installed. This simply could be two bolts, one on each side of the five-timber frame and hinged to a small super structure up from the two runners. The result would be a platform that could tip backwards conveniently, except a metal latch or plate would then be installed to prevent the platform from tipping backwards. The written description makes it much more confusing than needed, so I have sketched a typical stooker in the following figure.

This contraption was then hitched or attached to the rear of a baler; that in turn was hitched to a pulling tractor. I have stood on the narrow platform of a stooker all day, maybe for 15 hours total, and it was not uncomfortable, busy but not uncomfortable. One was only maybe 20 centimeters above the ground, so falling off was like stepping down one stair step.

So here was the process: One person drove a tractor and monitored the baler and its functioning, and another person stood on the stooker, grabbing one bale at a time from the baler rear chute and placing it on the stooker platform. When I did this, I would put the first bale on the extreme right, when facing the baler or to the front of the stooker. A second bale would go next to the first, a third next to that one, and the fourth finished off the first layer. The next bale would be placed on top of the first and second bales that had been put on the stooker already, and so on. When the tenth bale was placed on top to form the gabled stook, the metal plate was rotated to release the platform and with 10 bales of hay on that platform it naturally tilted easily back, allowing the full stook to slide off the rear and rest quietly on the ground. The platform, since it was pivoting about a point nearly in its center--and here was where one either used a previous stooker as a model or used some physics, or static analysis--would easily return to its horizontal position. The metal plate was rotated back to latch the platform for the next round. All this had to continue, as sure enough another bale was coming out of the baler chute, ready to be placed on the stooker for the next stook. As with all languages, there comes a point where one realizes that words are not the meaning. The stooker was the machine, or hay sled, but if one was on the stooker doing the stacking, they too were a stooker. So what was what depended on the context, as with everything.

I built a stooker that served us well for our time in the country. And now was coming the haying season. Fred Pilsworth had a field of hay ready to make by the first week of August, so during that week we cut about 10 acres of hay with our 5-foot hay mower. I wanted to use the horses and would do so later on our property, but on a neighbor's property I was still a bit reluctant. So we used the John Deere B for mowing. By August 10 we baled and stacked some 1,200 bales of hay. The weather was good for making hay, although we did have a few showers to slow the drying down. And we had another incident to slow us down.

There were many honeybees in the country, and there was a large hive on Pilsworth's property, back in the woods but there. And the bees would move around as much as a kilometer or more from the hive. Beekeepers know the range of honeybees, but they know that the bees are always in motion. During August they must be getting the biological and logical calls to get working on making honey to tide them over for the winter. That too was a stumble for me; many, probably the most of them, purchased a hive in the springtime and took all the honey from the busy bees. That meant that all the swarm of bees would perish as soon as cold weather set in. With just a small cut back on greed, a swarm of bees could winter over with a small portion of their earnings. It seems that most everyone is finally not good. I have been commenting on the big business practices and how their methods are evil, but even at the small-time level, one beekeeper seems to be blind to and uncaring for their fellow companions on the earth.

So Kurt and I were doing most of the haying at Pilsworth's farm, and one day I believe I was driving the John Deere A with the baler behind and our flatbed wagon behind the baler. Kurt was standing on the wagon and suddenly gave out with a yell. I turned around and saw that he had his right hand on his head. I stopped the tractor, and he told me that he had gotten stung be a bee on his temple. It took but another minute or so before I could see that he was starting to swell up. His head was red, and even his neck was puffing up. We were not far away from Pilsworth's house. We went there, and Fred and Irene, one or both, told us to go to the hospital, as a sting to the head is dangerous. I ran to our house, got the pickup truck and hurried back to Pilsworth's. Joy had her hands full with washing, cooking or housecleaning, but she came over with me to see if Kurt was all right. Well, he wasn't, but we got him in the truck and headed to Dawson Creek. I remember Joy commenting that he looked like the Incredible Hulk, a contemporary television program at the time involving a man who transformed occasionally into a large, balloon-like hulk. Of course this was serious, and we all seemed to know that. I remember driving as fast as I could under the circumstances of a dirt road to the border, but I concentrated on getting Kurt to the hospital. As we passed Baytree, Kurt commented that he didn't think he would make it, implying that he was having trouble breathing, and I am sure the pain was intense. He kept swelling up more and more. And we arrived at the emergency ward of the Dawson Creek hospital as quickly as we had ever made it. We had this sort of event twice before, Joy's birthing and my finger cutting, so we were accustomed to the procedures. And here there was no questioning of whether we had enough insurance coverage or ability to pay like would have occurred in the US, but they quickly administered to Kurt. He was given some antitoxins, I am sure, and maybe some antibiotics and was kept overnight. The following day he was recovering, and we were all so relieved when he came home in the afternoon. We would watch out as best we could to avoid bees and wasps, but one must continue and endure.

Around August 11 Fred sold part of his portion of the hay crop, 250 bales, to a hay buyer. They brought in a flatbed semi and loaded it up and went off. The following day I too had made arrangements for selling a load of hay, and another flatbed semi came and took 250 bales from my crop. I was told by the driver that he didn't have the money on him but would send a check when he delivered the hay. I should have known not to let him take the hay without payment, but I seem to have always been a slow learner, particularly when it comes to people relations. Two days later Fred told me that the hay was lost in the Pine Pass of British Columbia. An early snowstorm took the truck and load over a cliff. It sounds odd, or maybe dishonest is a better word, but I knew then that I had lost $500 of hay because I was not careful enough. And we could have used that cash right then.

Well, we had enough to do with more hay jobs and preparing for the upcoming winter. On August 8 or so George Fitzpatrick came over, and we became better acquainted. He had a thick field of clover that he had swathed and wanted to bale the hay. He had a swather and an Allis Chalmers tractor, but he needed someone with a rake, baler and stooker to help him do the haying. That was what I had, and this turned out to be a nice combination for George and myself. I had just gotten stung by some scheisters with the hay loss, and George was to redeem that episode.

It all was rather unbelievable but true. George swathed his crop, about 25 acres, and when we were discussing the baling operation, he commented that the hay was really heavy and entangled. His swather had put the hay in a continuous swath or windrow, and the hay dried out reasonably well. We were having nice sunny, long days, so baling could proceed normally. It turned out that my John Deere A would not have been able to handle the baling as the power takeoff that powered the baler was on the same clutch mechanism as the tractor. So I had been using the A and getting used to it for light hay. I could put the tractor in neutral, idle the baler and bring the plunger and rakes and whatnots up to some speed, give them some momentum, and then release the power takeoff, declutch it, shift the tractor into drive, and then clutch once more to take off with the baler now moving and baling. I didn't have a live power takeoff, and this was adequate with light hay, but George had heavy hay and from the beginning said his tractor, with a live power takeoff, could do it better than mine. He was right on.

On August 13 we began, he driving his Allis Chalmers and me being the stooker. It was a nice warm day, sunny, but at that latitude the sun hung at maybe 45 degrees and did not beat down so intense. It was fresh air, quiet surroundings except for the purr of the Allis Chalmers and the periodic chunk of the plunger of the baler. I wrote that the temperature reached 90 degrees F during the afternoon and was in the 80s most all day. We baled 491 bales starting just after lunch, and that was rather good. I had handled every one of those bales, and there were 49 stooks in his field. We had only just begun.

The next day we started in the late morning and baled until late afternoon. We made 681 bales, another 68 stooks, and still we had not finished half or even a third of the field. This was surely the thickest hay I ever experienced. The clover was dense, interlocking, and there was just a continuous flow of bales coming out of the baler. I remember standing on the stooker, removing a stook and placing it on the stooker platform, turning around quickly, and there was the next bale ready. George often had to take the tractor out of gear and let the baler catch up. But it was high quality hay, nice and dry, and it had a good aroma to it. Sweet clover is the right name for it.

In two days we had done 1,172 bales of hay, and much more was to come. As with many activities, once one gets going there seems to be an impetus, a momentum to go just a bit faster and set a goal of finishing the job. The third day we were in that sort of mode. The baler was going good, outside of needing baler twine almost continuously, but it just kept chunking out the bales. And George's tractor seemed to be holding up nicely too. So the third day we made 1,342 bales, starting in the early morning and finishing sometime in the late afternoon. I remember that 1,342 bales seemed like some sort of a record. But I was feeling good, and my muscles were limber, yet.

The fourth day we could see the end in sight. Remember now that this was about 25 acres of a continuous thick strand of hay with the windrows close side by side. By the fourth day we still had at least at least another 1,400 bales or so to make. As it turned out we baled 1,712 bales that fourth day, starting early in the morning and continuing until in the evening. This was good work, and George appreciated my efforts. I appreciated his opportunity to do the baling. He paid me $930 for the work, 25 cents per bale for using my baler, stooker and my efforts. I was pleased, and George got a large supply of hay. He was to sell much of it, and with it all stooked in his field, it must have been a comforting thing for him.

And about August 22 I cut a five-acre patch at Marvin Meyers's. I took the John Deere B over with our trusty mower, and the field was behind the house. During the day I found that I needed some kerosene for the B, so I went up to Marvin's house to call Fred or Irene Pilsworth who could then relay my message to bring up some kerosene. I knocked on the door--the house was a doublewide mobile home--and Marvin's wife greeted me. She had a housecoat on, and when I asked if I could use her telephone, she was very accommodating. I could tell that she was alone, and she took me to the phone, which was kind of sandwiched in a hallway, and she stood there next to me. I suddenly had a feeling that hadn't come over me for a long time, like I was attractive, and I could see that she was very attractive. I also noticed that the housecoat was partly opened, and I am not sure that she had anything on under that coat. I was sweaty and dirty and was still on my mission to get things done. She had, however, made me reflect, take notice, and I often wonder what if I had or she had come closer, brushed up against me or whatever.

With the experience at George Fitzpatrick's, I could see that we needed a larger tractor to pull the baler. Our John Deere A was just a marginal machine, and the B was too small to be expected to drive a baler with a wagon or even a stooker behind it. So on August 25 we went to Dawson Creek and bought a John Deere D at the Patterson Auction Mart. Somehow I was buying John Deere two-cylinder tractors. My very first tractor, the John Deere B, was a simple two-cylinder, flywheel crank start, with about 20 horsepower rating. It was dependable and did mostly all that I asked of it. The John Deere A was a mechanic's nightmare when we bought it at an auction sale for $100 in Ohio. I should not have bought it because it was worn out. But we rebuilt it, and it became a dependable tractor as well. It probably could give 30 horsepower and like the B was a flywheel crank-start machine. Both tractors had a small gas tank for starting and a larger 8 or 9 gallon tank for "petroleum distillates," meaning kerosene. Until the airlines began using gas turbine engines, kerosene was cheaper than gasoline, but that changed when gas turbines, which use kerosene as their fuel, became ubiquitous. Kerosene suddenly was more expensive than gasoline, and I found that gasoline also did not give the tractors as much power as kerosene, once the tractors were started. But somehow I just felt that a crank-start machine was more dependable and longlasting, and they were not so popular. That meant they could be purchased at lower cost and still deliver power. The John Deere D at the Patterson Auction Mart was a large machine, 501 cubic inches of swept volume in two cylinders, but it was sitting there in a row of other tractors kind of forlorn. It would provide all the power that I would ever want or need, yet it acted like it was dead. It certainly was ignored, and for $350 I just bought it. All the other tractors were twice, three, ten times what this one was priced at, and they had all the potential problems that this D didn't have: battery, starter motor, distributor, fuel pump, water pump, fan belts and more. The D was just another flywheel crank-start machine, and we found out that it started and ran as good as any of the machines there. I was a bit reluctant at first because it was so big. How would it start at -50 degrees C? It did have a power takeoff but not live, and would it still be able to pull a baler just getting ready to gulp a bale from start? I bought it with some encouragement from Kurt, who knew it was a good buy.

On the 27th of August Kurt and I baled 311 bales of hay at Meyers's, stacking them in a square stack on the east side of Twelve Mile Road just east of Meyers's. The pile was just to the north of a grove of popular trees and in a flat field--they all were flat north of Highway 49 for 12 miles or so--and that was where they would be until the winter when I would use my horses to haul them to our place. We now had a good supply of hay, roughly 1,000 bales, that I could sell and still enough for the horses and cow. And the summer was wearing on and out.

Yet, three days later, on August 30 we made about eight hectares of hay on a square plot of ground where we had stacked the hay from Marvin Meyers. This was Willi Meyers's land, and he told us to go ahead and make that hay, if we wanted it. He didn't want any of the hay and told us we could just have it. So again we had two days of work cutting the hay one day and baling another day. We were finished by September 4 and had another 300 bales or so to put in the stack right there.

CHAPTER NINETEEN
Threshing

As August was wearing away, the hint of winter was often in the air, particularly in the early morning. The air was getting crisp and clear, more so than in the summer warm days when the air is clear but with usually a tint of mist. It was now getting dry. We had long passed the summer solstice, June 21, and the daylight hours were starting to shorten. The ground was still warm from the summer, but at night one could feel that the seasons were gradually changing. Most of the hay had now been made in the region. I was looking at a few road shoulders near our farm that might have a few bales of hay, but there was also a host of other things needing our attention. One day the girls and Joy picked Saskatoon berries in British Columbia, about two hours away near the mountains. Saskatoon berries were a blueberry-like berry, somewhat bigger but not quite as sweet. They would be good in the depths of winter and even now when they were fresh. And the garden too was coming along nicely. We did have some potatoes and a few cabbages, lots of turnips and also some lettuce and carrots. The next summer would give us a better garden, but we were thankful for what we had.

The highway was very dusty now, and we lost our dear friend, Friendly, to the highway. There was not that much traffic on the road, but somehow she was at Pilsworth's and trying to cross to our place. She probably saw one truck coming and another truck came from the other direction and killed her. Of course the driver didn't stop, why bother with "just a dog." And somehow this was exactly what was going on in Ohio when we left. We gave Friendly a good burial and know that she is in a good spot near the good dugout. Her good friend, Prissy, was still with us, but she must have missed Friendly immensely. Prissy would have a litter of pups in the spring of 1980 which probably gave her some comfort.

I was now getting ready to do the oats and wheat. I had the binder in good shape and was ready to go. I remember people saying that you watch your neighbors, and when they start doing something, it is probably a good sign for you to do that. No one was yet combining grain, and my grain all looked like it could go another week or so. There was a bit of a breather now, so I began building one of the two granaries I wanted for the winter. I thought it would be a good idea to have a granary down by the road, where I planned to do the thrashing. That was in the southeast corner of the field, next to a low spot and a willow grove. I had plenty of 2 x 6 and 2 x 8 lumber for floor joists and wall frames. Also, I had a few 6 x 6 timbers for under the floor as the main lengthwise floor supports, the floor beams. These were a bit longer than the floor would be and with some large holes to accommodate a cable or heavy rope and which one could then drag the building some distance. So the beams would be sled runner, kind of. Floor joists would then be fastened crosswise to the two beams, the beams being about 6 feet apart. I had purchased a dozen 4 x 8 sheets of plywood for the floor of two granaries. I used six of them for the first granary. To economize, because plywood even then was expensive and I did have some lumber, I used 1-inch-thick boards for the granary walls and roof. Over the boards on the roof I used tarpaper shingles, somewhat against my wants, but it was quick, eh? I had just one main door on one of the gabled ends with sufficient access for shoveling grain into or out of the granary. I also elaborated a bit with some panels at the door so the grain wouldn't run out the door if it were to be opened and grain was against the door. I think this was about a 50-centimeter-high panel, and it was removable. That granary, 12 feet by 16 feet, was finished and ready for thrashing before I cut the wheat and oats.

On September 17 and 18 we cut the wheat and the oats. Kurt got the whole mess going and checked it out. He was the principle mechanic and seemed to know what was happening even when he had never done it before. I did most of the binding, though, riding on the seat behind the two horses who seemed to understand, like Kurt, how to do it much better than I did. It was a case of just keeping the cutter bar at the edge of the previously cut and bound grain and yet close enough to get a good swath. The binder would cut the grain at its base, the steams and grain falling on the pickup canvas after being pushed there by the classic reel that had four horizontal paddles. The grain and stems (straw) would then be transferred to another canvas that elevated the grain to a point about shoulder high, and it would kind of fall down into a vertical pocket where a twine was wrapped around the bundle. It was then pushed onto a cradle hanging over the side. When maybe 10 or 12 bundles were in the cradle I flipped a lever, I think it was a foot pedal actually, that dumped that pile of bundles of grain unto the ground, to be stooked. We spent two days binding the grain, and then we had to stook it all.

On September 19 we had our first significant frost, because it was 28 degrees F in the early morning. We may have had some earlier frosts but not enough to notice; this frost was noticeable, and it was white until past breakfast. It was fast approaching the autumnal equinox, September 21, when there was to be 12 hours of daylight and 12 hours of night. The next equinox would be March 21 of the following year, and from here on the daylight was less than the night.

Then on September 20 it warmed up again, and we proceeded to stook the wheat and some of the oats. It got too hot in the late afternoon, so we stopped and finished it the following day. It was a satisfying sight to see about 25 acres of grain in stooks, waiting to be thrashed. And while in the spring when I could see the trees greening up for summer, now it was a case of seeing trees' leaves turning deep yellow and the tamaracks turning brown. It would now only be another two weeks at most for the tress to hang on to their leaves and needles.

I needed to get a thrashing machine. I had committed now to doing the grain without the neighbors' swather and combine, and I began to scout around for a machine. I was surely fortunate as these machines were now obsolete and only the right person would have kept a machine. I saw many old thrashers, or separators, in Ohio and Indiana. This country seemed to be about 35 years behind, as the thrashing in Ohio and Indiana was nearly all gone by 1950 and everyone there was in a hurry to get a combine so they could work less and make more money. Really??? And the climate south of here was not kind to equipment or anything left outside. Many a separator was left outside, and the first things to go were the paper pulleys hanging out both sides of the machine and the wooden separator screen inside the machine. To be pounded by rain, snow and sleet rotted these away quickly. A good, operable separator would be ruined within 10 years left outside, and that was being optimistic. So I was able to find a good machine parked at the north end of Twelve Mile Road, right at the corner where it teed into an east-west road. It was owned by Ted Lefevre, and it was pretty much love at first sight. Any good mechanical engineer would see a separator and know that it had to be beautiful. Oh well, a separator is only one step down from a steam railroad engine, something like a 10-wheeler. These machines are big and show their intimate parts. It looks like a 10-meter-long box, maybe 3 meters high and 2 meters wide. The early machines would have been all wood, frames and panels, all except the cylinder. So it was a big oblong box and on both sides were protruding various diameter pulleys, mostly paper pulleys because they held up better than steel ones would. And they could be repaired easier too. Toward the front of the machine was a large-diameter drive pulley, this one of cast iron, that actually made everything go as long as it had a drive belt wrapped around it once. The other pulleys served crucial purposes and were connected with belts to one another. There may be as many as 15 pulleys and 10 belts moving when the machine was doing its job of separating the grain from its chaff or stems. The main drive pulley was usually directly connected with a shaft to a cylinder of about 1 meter diameter crosswise in the machine. The cylinder would rotate at a high speed and would be partially surrounded by a screen. Bundles of grain would be placed on a conveyor that extended out the front, usually portable so it could be removed. But the grain bundle would then enter the front of the machine and when you saw this machine in operation, you could see two claws grab the bundle and drag it into the chamber where the cylinder would whack it to pieces. Actually, the cylinder would pound the bundle, thereby getting the grain out of its husk and away from the chaff or stem. The grain could fall through the screen surrounding the cylinder onto another shaking screen and completely be separated from its husks. This was the main purpose of the thrashing machine or separator, and even today with all the glitzy new machines, the same process occurs. Now they have all the pulleys and belts and conveyors covered up, but it is still the same. Everything else about the machine and the various pulleys conveyors and belts were used to collect grain into a dumping bin, blow the chaff out into a pile next to the separator, and keep the machine blown clean. This machine was a McCormick Dearing, the classic of separators, although Case would argue that point. It had a complete set of belts, and they looked good, not rotted or worn out. And there was even a long drive belt, a continuous belt that could extend out at least 10 meters way from the drive pulley. It was about 30 centimeters wide, and it too was in good shape. I paid $100 for it, and to me that was a bargain. It had steel wheels, so I had to bring it home slowly, and 12 miles at 5 mph meant I must have spent more than two hours to bring it home.

Now we surely were ready for thrashing, as long as the weather held. On October 8 we moved the leanto that had come with the house and which we used the first winter for the same purpose as an entry room to the house. We moved this building, which was really a three-sided structure, to a position where I planned to build a barn. This would be the beginning of that barn. While I was on a roll with that project, we framed up a 7-meter-long by 7-meter-wide barn kind of attached to the leanto building. This would be the next oncoming project, to close in and finish the barn. But first we needed to get our oats and wheat thrashed.

On October 18, a Thursday by my notebook, we began thrashing. It was a bit problematic as it snowed in the early morning, and it was 28 degrees F, a bit chilly. By noon the ground was white, but we went ahead and started thrashing around 10 a.m. Jess Peachey and Willard Stuzman came over to help. I think they wanted to do this as they had grown up with thrashing and this would remind them of those times. Thrashing involves many hands for it to run smoothly, so with Jesse,

Willard, Kurt, Lori and Tim, we were able to get a good start, even with the snow. I remember Jess bring a wagonload of bundles in from the field and driving that with the horses up to the separator, right by the side of the lead conveyor and with a pitch fork (the three tiners worked best) toss the bundles one at a time onto the conveyor, parallel to the machine. I think we tossed the heads in first so the grain in its husk would get hit by the rotating cylinder first. There was a lot of dust, vibrations, noise and the unique aroma of fresh grain, hard to describe.

The machine was pulled by, or powered by, the John Deere D, and it just chugged along nicely. Some people I had talked with before we did this suggested a bigger tractor or something other than a two-cylinder engine so it would be smoother. The D didn't seem to have any problems, and I think it must have enjoyed today's work. I heard no complaints from the tractor. We ended up the day's work with about 110 bushels of oats and also about 70 bushels of wheat from an acre and a half of wheat. That made me feel good as that meant our yield from the wheat was nearly 50 bushels per acre, not bad at all. And it was good clean wheat; it would be good for baking and cooking in the coming months. We stored it in a couple of barrels and kept it in the log granary, away from the horses. The thrashing did go a bit rough because of the fresh snow and resulting moisture, but everyone had a safe and enjoyable time of it, I think. We still had another 10 acres or so of oats to thrash. And we had to wait for that because the weather took a turn for the cold. The temperatures hovered in the negative single-digit Celsius range for the next few days, and we had ice forming on the dugouts. I think the ice was actually thick enough for skating, but that did not occur yet.

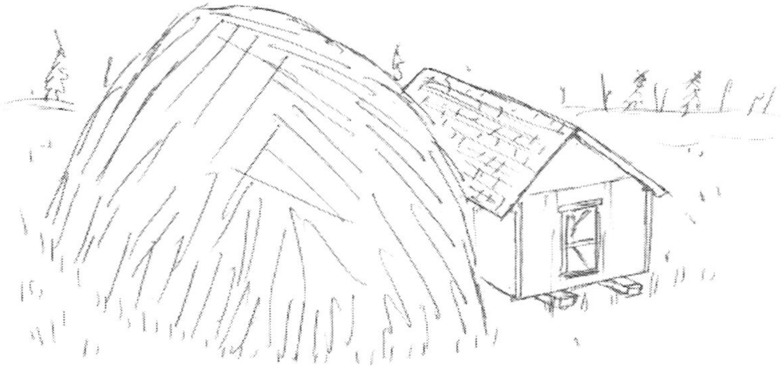

On October 29 the weather was a bit warmer, in the plus single-digit Celsius range, and we finished up the thrashing ending with about 750 bushels of oats. We put the bulk of it in the new granary and took a small portion up to the barn for the cow and horses. Besides the nice harvest of oats and wheat, we had a large pile of clean straw next to the new granary in the field. It was like a dome, maybe 6 meters high and that much in diameter. We would use the straw as bedding for the horses, the cow, the chickens, and the pigs when we got some. I had never directed or even participated in an actual thrashing of grain, and this was an experience I cherish and will remember forever. I often wonder why the farmers were so ready to give up the event of thrashing, or for that matter, the events of grain stooking, corn shocking and shredding, making loose hay, using horses that were alive and thinking companions on a farm operation, and using open pollinating corns and other grains. I know, I know, the farmers couldn't afford to not go to tractors, combines, and all the other paraphernalia of the modern agribusiness. That is another example of telling a big enough lie that people will believe you or believe the businesses that were finally the only ones who actually made profits from the modern methods.

And during our time in the Peace River Country we were faithful guests of the Baytree Mennonite Church. Joy and I started taking some "instruction" classes, and I think back now that that scenario was the beginning of the rupture with that church for us, surely me. I couldn't understand why they knew things that related to anything that would make me qualified to be a good person. I had this feeling that their God was dead, as I wrote earlier, and it is, I think, a sad commentary that groups need to indoctrinate potential new members. Anyway, for some time I just wanted to be humble and meek and go along with the exercise. Kurt and Lori were going to school at the Mennonite school, so that kind of qualified them for membership. They did join the church on November 3, I think. That is a Saturday by my count, but I suppose a Saturday should work for baptism in a church as well as a Sunday. As I write this, that day was cold, 7 degrees F in the morning, but clear and sunny. I can imagine it being a bit of snow cover, the trees bare, and the air crisp. Significantly, Tim also could have become member, although he was a bit young. He did not join, and Joy and I did not join either. By this time there were a lot of little petty and odd statements, behaviors and incidents that turned me off, things that are not necessary for this recital, epistle or story.

A problem seemed to be brewing up at the school as well. Carol Peachey was the full-time teacher, and our children all seemed to be accepting of her. Kurt was doing his correspondence course work to finish a high school program. Lori was now becoming a real grown young lady and I think studying the equivalent of sophomore high school work. Carol seemed to be helping her along on that even though that was beyond her actual formal school. Tim was about an eighth-grade equivalent student and was the model student. He always enjoyed book reading and studying in general, so he had no problem with Carol or the school. Heidi was maybe an equivalent of the sixth grader, and sweet Heidi was also a good student, not one to cause trouble. And David Donaldson had joined the congregation, somewhat like us in that I don't think he was a church member, not a bonafide Mennonite. His son, Hilding, went to the school, and somehow David (I think) decided that he would help out at school. He actually began his "helping" in late September, and by early November he had made enough inroads that he somehow decided that he was a part of the school faculty and was entitled to mete out discipline, if need be. And for some people there was always a need be. Well, our children kept us informed of some of the happenings going on, and while they all four were not the type to react against unwarranted scoldings or even harsher disciplinary measures, we saw that it was not good. We did talk with Carol about this, and I think she thought it was not anything to be concerned about and that she, not David, was still the teacher. By the middle of November it was becoming clear that our children were getting uncomfortable at the Baytree Mennonite school and that something had to happen. We finally decided, Joy and I, that we would keep the children in the school through the year, letting them participate in the Christmas program, and in January 1980 we would send Lori, Tim and Heidi to the Bonanza school, taking them out of the Mennonite school.

And something else came forward. Maybe it always was there, but Joy and I had worked hard, sacrificing maybe too and trying to do what we thought was right. It all finally begins to take a toll because a marriage must be worked at. There surely are no true answers, and I believe that acceptance of another person is the key; giving them enough room and respect is the essence of any good relationship. It is not having exhilarating sex or highs, not accumulating money or possessions, not even having insightful conversations with another person. It is just taking others at their face value and respecting that face value. That has been hard for me, and I know that others too find that hard to the point that they kill their God so they don't have to do it. Somehow I think it will all come out right in the end, as I describe Heaven in the appendix. So we were essentially avoiding each other, Joy and I. Oh, we sat at each meal together, even had conversations and made plans together, and even expressed our love for each other once or twice. Not often. But I know I was struggling with the knowledge that I had no dependable income, and while I could see a clear goal of becoming self sufficient here in the Peace Country, it was a long way off, and I just became rather down about not being able to fully provide for the family. So I began to seek some outside employment to get a dependable income. After all, the official reason that we were acceptable for immigration by the Canadian was my education and skills, not my monetary wealth. I was a registered professional engineer and could design nearly anything pertaining to mechanical engineering. I could teach or do expert witness as well. So I gave it a go. Probably not full force, but I did go to Dawson Creek, a printer, and had a box of business cards made up. I still have some of them. It was a single first step or crawl. I then put some cards in the Provincial offices where likely customers would be for environmental control design (heating, ventilating and air conditioning, HVAC). I put an advertisement in the Provincial paper at Spirit River and in Dawson Creek. I went to the Northern Lights Community College in Dawson Creek; I could teach physics or chemistry if need be, certainly mathematics. I got not one nibble. I went to various machine shops and repair shops in Dawson Creek and Spirit River, even went to Grande Prairie in response to an ad seeking an engineer for design. Nothing. And it was just like I had thought and read about and discussed with others: One can get too educated. That is nonsense of course, but business people, and anyone, are suspicious of someone who may have more education than they have. Up here it was just very clear, with the exception that they were friendly and helpful and good neighbors. But this exercise in attempting to join some engineering group where I could utilize my skills and education finally left me at the farm. I had started it going alone, with the family, and now I did have to go it alone. Somehow we had to get the right mix of produce or products to sell to manage a sustainable income and do it the right way. We did have a good start, I believed, and still do. But times were getting more complicated for us.

Joy had been more aggressive about money matters than I was. She began to work as a housekeeper toward the end of October at Ralph Tompkins's. It was an arrangement where she would work from about 8 a.m. until 4 p.m., maybe eight hours, and her earnings were really helpful for the family. I think Joy enjoyed doing the housework there because it was heated with a central furnace, probably oil heat, and was just about like a typical ranch-style house that could be in Ohio or Indiana, kind of going back for her. She would have Sunshine and Charity with her, and it was all nice and comfortable. She also had the luxury, if that is the right word, to return home in the evening to a warm house where often Lori and Heidi had a nice delicious supper prepared and ready. Joy could then relax, but we were still struggling to get along, and in my own selfish slant on it, it now was a case where Joy was the breadwinner.

Ralph Tompkins had a farm about four kilometers directly north of Bonanza, so it was a bit of a drive for her. Ralph, however, was not a farmer. He did, I think, inherit the land from his parents, but he was an oil man. He had a roadgrader, a bulldozer, some dump trucks and maybe a backhoe. He contracted to clear a site where a potential oil rig could be placed for drilling oil. It seemed to be quite lucrative, and he had a comfortable house and a set of outbuildings. One of his outbuildings was used for repairing his equipment, and he offered to let us come up and work on our pickup truck when we had time. We were having problems with the truck as it was getting difficult to start, and when it did it would sometimes sputter and sometimes even die right on the highway. It seemed to need a good tuneup which we planned to do at some time. The one-ton truck was dependable and was our major vehicle for the time, although Joy used the pickup for a while to drive back and forth to Tompkins's.

We received a letter from Joy's parents indicating that someone was interested in our young stallion, Chief. We quickly returned a letter asking how much Bill (Joy's dad) thought we ought to ask. He had been handling the horse since we left, and he knew its disposition and potentials. He corresponded with us again, and we sold Chief for $1,500 officially on November 1, 1979. I think the man came a couple of weeks later. Joy's dad must have had an empty spot at that time. I sure did.

November 7 was zero Fahrenheit at sunrise and clear, and it continued cold for a week or more. Winter was now upon us. Then around November 10 we got a Chinook. And it lasted for about eight days, letting us built and finish a second granary that we then placed against the house, where the old leanto had been. This would have to serve for the second winter as the snow room. And with that respite, we advanced on the barn, putting some siding on the walls and nearly closing in the granary area, that area adjacent to the old leanto and extending out a bit. We did put the roof on the barn, kind of cobbled together. It was lathing or slabs laid on purlins made of poplar poles. This would be covered with black plastic and more slabs on top to keep the plastic down. All this took time, so I had to finish it before the winter set in good. The barn was not really a clean building, but I rationalized that it was just a temporary barn and would shed rain and snow, and provide shelter for the horses and cow. There was also a good space left over to accommodate a few pigs. We would see.

On November 14 the temperature reached 60 degrees F and the ice was melting on the dugout. Since we had maybe 6 inches of ice to begin with, it could melt a long while before it would be too thin to walk on. And it did have some snow cover which slowed the thawing.

As I was to write in our diary, "*it has been a very nice summer and fall-hope the winter is good also.*" Flossy was doing well, giving about two gallons of milk every day, hand-milked by the way. And her young calf also was healthy and frisky. The calf would get some of the milk, and we would be using at least a gallon every day for the six children. And Bess, Morning Star and Prince were doing good. I think they were ready for some more work, but I am also sure they were enjoying the lounging around and soaking up the warm sun.

The next day we were back in a deep freeze. Not exceptionally cold, but like the year before, cool at maybe 20 degrees F. And we were coming up to a major event. Sunshine would be one year old on December 4. How time flew by. She was crawling around, talking a bit, had three teeth, and was just what a parent could want: healthy and active, and beautiful, like her siblings. We all celebrated her birthday together at home. And as if it were a bonus for us, we had another Chinook around December 9 which lasted a few days. I went at the barn roof with a sense of mission on that exact day, just realizing that even though the temperatures in the morning were -20 degrees C, I needed to get something substantial on the roof. I can remember working on the roof with enough clothes to stay warm and a warm pair of gloves, not mittens this time. As we, Kurt, Tim and I, moved around we started taking first our coats off, then our gloves, and before noon we were perspiring. We were working hard, but the Chinook had come again, and when it is 60 degrees F, one can work comfortably. We installed a big black plastic tarp material over the slabs, and then more slabs went on top. That finished the barn roof, and we had previously installed stalls, so we brought the horses and Flossy into the barn for the first time. Flossy didn't trust the barn as she had gotten accustomed to the log granary; it was home, and this was just too much. But she adapted nicely and didn't drop her milk production.

About this time Joy told us that Ralph Tompkins had offered to let Joy use his four-door sedan, a Pontiac, I think. It would be more dependable than the pickup truck and more comfortable. So from that time on Joy was driving the four-door sedan. It seemed to me to be driving another wedge between Joy and me, but then I was getting sensitive and defensive. Maybe I had been that way for a long time and Joy was just reacting to me.

Anyway, our chickens were doing well in the up and down weather too, and we were getting a fair supply of eggs. The Pilsworths had chickens too, and they almost had a science of having just the right number of chickens in the coop so they wouldn't get hot or cold during the winter. We didn't have that luxury, but we would find out if our egg supply would dwindle during the really cold stretches. During the

Chinook, one day someone noticed a hawk circling overhead, and it then swooped down to get one of our chickens. I think he, or she, may have almost nabbed a chicken, but we ran outside and the hawk left. Surprising since we were in the bush, the woods, and the hawk had to have a sharp eye to spot the chickens and then a good guidance system to avoid trees and chicken fence and whatnot. Maybe we should have given the hawk a meal, a gift.

And as could be predicted, on December 11 the thermometer measured -25 degrees C in the morning. It would continue to drop every day then, and on December 15 it was -50 degrees F at 7 a.m., dark and cold. But we had about 30 centimeters of nice fluffy snow that dropped down intermittently during the previous three days. And during this time we were visited in the evening by a representative of either an oil company or the Provincial government. The man was friendly enough, but he obviously was on an official mission to extract something from us. Well, he told us that an exploration group wanted to run a seismic line through our property, somewhere near the cleared field but in the bush. I was acclimated to the situation in Canada. One owned the land, but the mineral rights were that of the Queen, the Province of Alberta in this case. So what could I do? Say, "No, I came 3,000 miles to get away from your type"? And I didn't need the oil. That would not fly and it would be quite ignorant. They did, I think, offer to pay us a stipend, maybe $250 to do it, so we once more had to compromise and give in. I felt it to be more of a surrender, but, whatever, they were to come a few days later. Middle December was past the freeze up, so the ground would be firm and their machinery could move through our property easily.

Joy was gone all day during the week, and I remember one day, shortly after the visit, that I was in the housekeeping mode, the stoves going and cleaning up some, when I heard an explosion and the windows rattled. It happened probably a dozen times, with a lull of maybe a half hour between shots. At first I was not understanding what was going on. But it all came to me: Schlumberger was running their seismic line through our property. It didn't damage the windows or anything, but it was surely an insult to me, even with the $250. And looking back, the real insult was to the earth. How would a person feel if some alien being or germ or virus entered our body and set off a minor explosions, measuring the effects of that shock and analyzing it to see if they could get something out of us? Oh, yes, the earth has no feelings or soul or all that stuff. It is here to rape, plunder and move over. I had hoped then that they would not find what they were after, but they still ripped a path, some 8 meters wide, cutting the trees down or pushing them over, from the road straight west to the muskeg and beyond. I think the next parcel of land west was Provincial land, so it was totally theirs. I walked down to the damaged earth a day later, and it was not pretty. Every 30 meters or so was a pile of earth where they had drilled, dropped an explosive charge down the hole and ignited it. At the impact and for a few minutes after there was a seismic sensor that recorded the vibration of the earth to get a rough idea whether there was oil or some deposit under the ground. As I said, they must have done this a dozen times on our half mile stretch of the line, and they kept going. Oh, Schlumberger (pronounced Schlumber-Jay) is a large exploratory company that has a unique truck, three-axled and designed to carry all the necessary equipment, instrumentations and personnel to conduct these events. They are making so much money that it is just beyond a normal person's comprehension.

And I had never anticipated this.

We had written to my father and Joy's parents in Indiana some, probably not enough, but we planned to travel down over Christmas of 1979 to visit them. That would be a nice break for us.

CHAPTER TWENTY
Preparations for the Second Winter

Our wood supply for the second winter was better than the first but still precarious. We just had so much going on that we never actually had a big pile of wood ready for the stoves. It was always a day-to-day operation, but it was coming.

The road crew or Provincial highway department too knew that winter was coming, and they must have planned to have Highway 49 completed by the next summer, so they went full blast in late October to get a cover coat of dirt over the dirt grading. It was a layer maybe 10 or 12 centimeters thick, and it looked like just more dirt, maybe with some sand or fine gravel, and it must have had a binder mixed in. It was applied much like I had seen paving crews put asphalt pavement down. And for a few days it was smooth and tractable. Actually, as long as the rains would hold off, that pavement was like an asphalt pavement. So the following summer the crew would come back and finish their road work.

I had spotted some logs cut and stacked up the road from Pilsworth, south of the corner. It looked like they may have been cut some time ago, maybe even two years before. They were the logs left for clearing an east-west path, a clearing for either a pipeline or anticipated power line. One could see that it extended far to the east and west. So I asked around and found out that the Province probably owned the ground and therefore the logs. I went to Spirit River to inquire and filled out a short form, purchased the pile for I think maybe $15. It cost more to drive to Spirit River than the logs cost. But it was then all on the up and up, and I planned to go up the hill with my horses and haul them back. It would be nice firewood if nothing else. In the middle of December John and Irene Trinier visited us, and we got to talking about the horses, as he too was interested in and appreciated horses, particularly draft horses. I suggested that maybe we could drive them up the hill south of Pilsworth and get a load of the logs I had purchased. He was ready for that, so we harnessed up Bess and Morning Star and hitched them to the green wagon. We had the side boards on, so we could only take a small load. That would work out just fine. We went up the hill and to the location where the logs were supposed to be. And wouldn't you know it, they were almost all gone. It really seemed ironic or worse; they had lain there for a long time and not two weeks after I spoke for them, someone hauled them off? It just didn't make any sense, but there were a few left, and we loaded them on the wagon and drove back down the hill to our place.

Looking back now, the country was inhabited by people eking out a living, and somehow someone had run across those logs and just took them, before someone else would. Everyone was grubbing it out and not looking around or back or frontwise to see if what was transpiring was legal or moral or ethical. Those terms are just way too abstract and impractical.

John and I had a good workout with the horses, and the horses too were happy for the work. And finally we did have a few firewood logs for the next time.

We had planned to travel to Indiana over Christmas, and that would mean we would have our house vacant for that time. Well, John and Irene offered to stay in the house then, keep it warm and take care of the cow, the horses and the chickens in exchange for staying in the house. That was a good bargain for us, and I hope that they thought the same. Of all the church people and neighbors, John and Irene seemed to be the most responsible and not distracted. Others just had too much going on. So we planned on having them at our house from around December 20 until January 2 or 3.

We were a bit better prepared for the second winter. In particular, occupying the house all the time and having heat in it meant that the waterline that had given us trouble during the first winter did not freeze up. It was to be expected, and we had good fresh water all the time now, no thawing the line with salt and straws. That was a major improvement for us.

During the week of December 15 to 20 for some reason Joy asked if I would go with her to Tompkins, just for a half day. So I did, and while we were there we got to talking about the cold weather, that was always on people's minds, even the natives. And it then came up that on Ralph's property was an old log house, to the far east of his property not too far off a side road. It was maybe the original building on his land, and one of his relatives, or maybe just a friend, lived there during the warm months. My details of this are not too good because I was now getting suspicious of Ralph and all his actions and conversations, besides the fact that he was an oil man.... Anyway, around December this man would move out and go down south, to the Okanagan or around Vancouver maybe, returning in the spring to spend the summer in the log house. Ralph took us to the building; it was unlocked and we went in. It was now middle December, it had been cold off and on for two and a half months, and when we went in there were live plants, mostly flowers but green plants around the room. It was a low-roofed and ceilinged building so the ceiling was maybe 6 feet high, 2 meters. A tall person would have had to duck to walk around the room. And the walls were authentic logs, like our house, and they seemed to be tight. The only windows were on the south facing wall, and they were overhung nicely and set back so the sun, from mid November to March, could easily enter the room, yet it was protected enough that at night the heatloss would be reduced. Ralph assured us that these plants would be alive when his friend or relative returned in May. I never asked, but I wondered if the plants needed some water during that long four- or five-month period of neglect. Actually, cool air and cool plants do not need much water, so surviving for four months or so without a shot of water was possible and seemed to be how they survived. I just couldn't imagine Ralph Tompkins watering some plants in a log house 500 meters from his house. There wasn't any money in that sort of exercise. But that was my bias. At any rate this was a good case study of how every place on the earth gets the same number of hours of daylight in a year, just distributed differently. And up here at 58 degrees north latitude, when the sun is just barely in the southern sky, a south facing window receives its maximum amount of solar energy. Another observation I made was how the building was tight, and as long as the door was not opened too much, the air inside the building would remain relatively warm and not drop to freezing or below. Our log house was not suited to this sort of condition, but it did provide good information for the years ahead when energy would become scarce and expensive, even wood.

As we were getting ready for our trip south, we decided that we should get a telephone. Going to the Pilsworths to make telephone calls was getting to be an imposition on them, although they never said that. I sometimes wondered if they enjoyed us coming over to make telephone calls so they could overhear and kind of gather some information on us. But at any rate, we did inquire about having a telephone installed, and it seemed that it was going to be a simple case of burying a line up our lane from the road and tying in at the road. And the frozen ground didn't seem to concern them, so sometime in December we had a new telephone. It was a wall model and installed against the northeast corner of the kitchen, just to the right of the cooking stove.

And while we were at that, it was decided that when we got back from Indiana we would get a television. The children would not be in the Mennonite school anymore, and our obedience to the theology of the church was slacking off significantly, so a television was OK.

The BayTree Mennonite school had their yearly Christmas program on December 18, and our four school-aged children participated in that. Carol Peachey and the rest of the church members were in no way holding the coming withdrawal of our children from the school against us. I think they probably understood the events better than I did. And I remember that the Christmas program was well done, again held in the one-room schoolhouse, and there were lots of good treats and drinks and fellowship to be had. It was a packed house, probably for the last time in that building. And I remember that since Joy had been working at Ralph Tompkins's his daughter started going to the school. So she and Ralph were there too. I spent very little time talking with Ralph. I did find out, from Joy, that Ralph's daughter was with his wife who died about two or three years before we moved there. He had met her in Utah, I think, and she may have been a Mormon. That doesn't have anything to do with it, but she then moved to northern Alberta to live on Ralph's farmstead. I don't know any of the particulars of what reason there was for her dying so young, but Ralph then was left with a daughter, probably 6 years old at the time. He had another live-in type woman, and I don't know any of the particulars about that either. But everything just seemed very different when it came to Ralph Tompkins, even though he was being very helpful in giving Joy a good housekeeping position.

The church had their plans well underway for building a new church and school on a corner of Jim Sadlier's farm. They all seemed anxious to get that started, and I think sometime in late February or March they did dig out for footers and make ready for pouring concrete as soon as the weather permitted. We were not in that loop anymore, so I only helped out once or twice with the construction of the new church, although we still went often to the Sunday church services.

During the middle of December I did go ahead and construct a pig pen in the north side of the barn, near the horse and cow stanchions. I had saved a number of black poplar logs and had them sawed into boards which I then used as a floor for the pigs, right on the ground. The barn was all dirt floor, no poured concrete or whatever else. And someone had told me that black poplar would last indefinitely as barn flooring, the manure and urine even making it stronger and more enduring. I thought I would give it a try and find out if, after 20 years, they were right. Well, during the course of my working on that detail of the barn, I had a visitor come and talk with me about the Watchtower and other items relating to the Jehovah's Witnesses. I wondered if our new neighbor, Roland Gauche, had perhaps suggested that they visit me. Whatever, we had a good talk, and I have no problems with the Jehovah's Witness people. Their arithmetic of about 20,000 people being saved and going to Heaven is more than odd and maybe racist and bigoted, but who isn't racist and bigoted? They do say that they are pacifistic, but when I pressed him on that he always backed off. One must resist in the name of defense, kind of making Jesus out to be a liar, eh? And he made no headway with me, nor me him. He took me from my work for about half an hour, but that was OK. I did get the pen finished and even hung a gate on the fence to let the pigs into the pen. It would be quite adequate when we purchased a sow, maybe one that was going to have some little pigs. But that would have to wait until we returned from Indiana.

We planned to drive to Shelby, Montana, and catch the Empire Builder, the Amtrak train that ran from Chicago to Seattle, Washington, daily. It would be a hard day's drive to Shelby, but our pickup would negotiate it good. And then Joy said Ralph had suggested that we take his car to Shelby; it would be more room and comfortable and safer, eh? So we took Ralph's Pontiac to Shelby. It was my idea to take the train. I couldn't see us driving to Indiana with any sort of vehicle. It would have been at least three nights each way in a motel, $400 or so, it would have been at least 2,500 miles at a dollar a mile, and that would then have cost us probably $5,800 for fuel and lodging plus another $400 for meals. The railroad was maybe $800 round trip for a coach seat for each of us, all seven. Sunshine would be on one of our laps. And it would be safer, quicker and, unless we drove breaking speed limits, just as good for sightseeing, and we would meet some interesting people. All that happened. But we first had to leave, and as I said, John and Irene Trinier stayed in our house for the time we were gone, from December 22 until January 2.

CHAPTER TWENTY-ONE
Return to Indiana

We drove south from our farm, heading east on Highway 49, even though it was still dirt. In the winter that was no problem and might even be better than some of the paved roads. We headed on as direct a route as we could to Shelby, Montana. The Empire Builder, the Amtrak train that we would take to Chicago, was scheduled to arrive in Shelby at 11:45 a.m., just before noon. So we made it a point to be there at the station on time, even though we knew it would probably be late. It was about 750 miles from our house to Shelby, so we left on December 21 to meet the train on December 22. We stayed overnight near Calgary, and that left us with a drive of about 100 miles to make it to Shelby by noon. This was the downside to traveling by train; it took a long time just to get to the train, so in reality, we drove 750 of the 2,500 miles (approximately) with the car, and that meant we only saved 1,750 miles by taking the train. So doing the economics again, we saved driving 3,500 miles (about a dollar a mile to be fair). We still saved about $700 on motel rooms and $300 for meals, so by driving we would have spent $4,500. Our tickets totaled $800 so we saved $3,700. And we enjoyed the train trip, relaxed and arrived quicker.

The train arrived nearly on time, I remember, and it was a snowy day, cloudy and dark. We headed east and got to see eastern Montana in the winter, on the rails. Winter in the upper plains is a very interesting sight with lots of snow, barren fields and, sometime off in the distance, a range of mountains. We passed through Havre (pronounced Haver by the folks, but I am always obstinate and call it Hav ray, right?). We passed by Malta, Glasgow, Wolf Point on an Indian reservation and entered North Dakota a half hour or so west of Williston. I kept thinking how all this land is really Indian land, just confiscated and raped by whitey. By the time we were in North Dakota it was dark and at Minot, North Dakota, a major station stop, it was late in the evening. Traveling across North Dakota at night was mostly just trying to get there through Rugby, Devils Lake, Grand Forks and Fargo. The coaches for the trip were comfortable, and Little Sunshine was about to get some good quality sleeping. Even Charity and Heidi had plenty of room. The rest of us had to make do, but it was like trying to sleep in a large overstuffed chair in a living room. And if one got stiff or needed a break, there was always the lounge car or just a walk around the train. I remember as we went through Minnesota, about an hour or so west of the Twin Cities, we were moving along next to a major highway, maybe US 10, a four laner, and we were going right past them with snow swirling around us and on the highway too. We must have been going 80 mph or a bit more. I since found out that legally trains cannot go faster than 79 mph, but that gets bent a bit, I know. We passed through the Twin Cities in the early morning and then traveled along the west side of the Mississippi River to LaCrosse, Wisconsin, kind of retracing our steps from a year and a half ago. We zipped through Wisconsin, through Milwaukee and arrived in Chicago around 5 p.m., just late enough that we had to spend the night in Chicago to catch the train the next day, the Cardinal, which left around two o'clock in the afternoon. So we spent a night in a high-rise motel in the Loop of Chicago, nothing particularly exciting. On December 24 we boarded the Cardinal, and at that time it took a route through Indiana that went through Logansport, Marion and Muncie before it also went through Richmond, Indiana, and on to Cincinnati, ending up in Washington, DC. We got off in Muncie and were picked up by Joy's parents. It was a relief to finally get to our destination, and we would spend Christmas and up to the New Year there. The weather, as I recall it, was cold and damp. Not as cold as in Canada, but it was more uncomfortable. Canadian cold was dry, even though it didn't seem like it when we were there. Mostly the temperatures were right around freezing, 32 degrees F or zero Celsius, and there was a bit of snow on the ground. We took it all in, and the children I think were glad to be back for a bit. We planned on staying at Joy's parents in their second house, the Wysong place. It was a big two-story, brick house, blocky. And it was old and interesting with lots of high quality wood for woodwork; the floors too were substantial, and all in all it was a comfortable place to stay for a few days. There were no plush carpets on the wood floors, no curtains in the windows, and the lights were mostly just utilitarian, nothing fancy. But we were grateful, I know I was, and we settled in.

My father, Opa, lived in the same community just about two miles away from Joy's parents, and I would spend a night or two with him at his home. I think maybe some of the children accompanied me there.

We had a good Christmas there, starting almost from the minute we arrived, on December 24 late in the afternoon, Christmas Eve. It seemed like everything was about the way it was when we left. I felt a bit out of place and was looking to do something. And one of the things I wanted to do was visit some of my good friends at the University of Dayton: Bob Mott and Jess Wilder. So one day I went down with my father, Opa, to meet with them. I had no real problem with that as I would have some good time with my father. He was still grappling with my move to Canada and just couldn't, or didn't want to, understand the whole venture. My grandfather, my opa, would have understood it very well. And we had a good visit with my friends, had a nice lunch at the University dining room and everything looked about normal to me. I had left the university with tenure and was told at every juncture that I was welcome to come back and get my tenure back. They were good people, and it was a strong pull on me to just hang it all up and come back to UD. That was tempting, but I was still committed. During our luncheon discussion, however, Jess Wilder mentioned that the University of Wisconsin was starting a mechanical engineering program somewhere in the southwest of the state and he had read the job description. He thought it sounded much like mechanical engineering technology we did at UD. That was good solid practical mechanical engineering and not fast and furious research. I thought about it and actually got an address of a contact person. It was a way out, and that all was on my mind, probably more than I realized. While there I also got another contact for Montana Tech in Butte, Montana, and that too sounded like good solid mechanical engineering teaching. Both the position in Wisconsin and the one in Montana were located in rather rural settings, at least more so than Dayton, which was the big reason I became so disenchanted with the Dayton area. There were just too many people in one spot, and the sociology and development of the cities seemed to accentuate the crowdedness and hum drum of congestion. I would just stay put in the Peace River Country.

While we were visiting in the region, I think we visited in Ohio, mostly with one or two families of the Old Order German Baptist Church. Our daughter Lori was good friends with twins, Donna and Doris Diehl, who were now members of that church and lived on Greise Road close by Bradford, Ohio. So I am sure that we spent some time there visiting with their family. I am sure also that the conversations were a bit tight and on the edge too. Joy had been a member of their church before we left Ohio, and she was a good friend of Ruth Diehl, the mother of the twins. Ruth was nice, meek person who went along with all the jargon and rules of the group. And she was always civil to us. I was always able to get along with the people down there but kept them at arm's length, so to speak. It is also possible that we spent some time with David Grim and his family, but that was probably rather a short visit. Our priority was to be with Joy's parents and my father as much as possible.

So we had an itinerary that took us to Fort Wayne, Indiana, to catch the Capitol Limited sometime in the very early morning of January 2, 1980, about 4 a.m. if I remember correctly. It must have been a cold, dark day. My father, Opa, drove us there. Actually I remember that he was getting on in years but thought he could still safely drive a car. He was the archetypical American driver; he felt it a right, almost an obligation, to drive his car. He felt that it was liberty, economy and comfort all at the same time. That is being proven every day to be the untruth of it; an automobile is a weapon if directed by an inattentive, incompetent or inebriated person, it is expensive--about one dollar per mile which must be more than a first class airline ticket--and riding for more than two hours or so leaves one emotionally and physically fatigued. Growing up with him I remember his adamant rejection of the railroad, so taking us to Fort Wayne to catch a train must have rubbed him raw. Plus we were going back to Canada, in the Godforsaken northern part of Alberta with nothing there. But Joy and he were not on good relations, and she did have some valid reasons for that; he did too. Anyway, early in the morning as we were getting ready to depart in Opa's car, he just normally got into the driver's eat, ready to drive his passengers to Fort Wayne. Well, Joy had told me to drive because she was uncomfortable with him behind the wheel. I thought that was a good idea as well, but I was not in a confrontational mode at all. We were leaving, and who knows when we would be back or even see my father. But Joy then, herself, confronted Opa and essentially demanded that I drive his car or we would find another way to go. She was using diplomatic brinkmanship, but it worked. My dad, I think, was not eager to make any sort of argumentative scene with six of his grandchildren around. So I drove to Fort Wayne, some 100 kilometers north of Winchester. It must have taken us an hour and a half, and it was a highway I had been on many times. We made it safely, and I know my father also returned home, on his own, safely. But this too was a kind of incident that I had gotten used to with Joy, and the children too had had some experiences along the same lines. I do remember then, and now, hoping that he would forgive me of the event and go on with his life in ease.

The train ride back to Shelby, Montana, was comfortable and faster than coming down. The Capitol Limited arrived in the morning at Union Station, Chicago, and we had a few hours to wait for the departure of the Empire Builder in the early afternoon, 2 p.m. or so. I think the weather was not exceptionally cold, but it was the depth of the winter, so not warm either. We retraced our travel down from Chicago on, traveling through Wisconsin and arriving in the Twin Cities around 10:30 in the evening. By 11 p.m. we were heading west through Saint Cloud, Little Falls, Staples and Detroit Lakes, Minnesota, before entering North Dakota at Fargo. Even on the train one could feel that the weather was cooling down, and outside there was more snow than in Indiana or Wisconsin. And being the prairies, one could just sense the lonely winds blowing so that by the next morning as we passed through Rugby, North Dakota, and approached Minot, it was definitely different country than Indiana and Ohio. I actually felt that we were heading home, and we were. It would take us until around 4 p.m. to arrive at Shelby. In January in Shelby, that meant it was getting dark, so we found the auto and headed north without delay. I think we may have stayed overnight between Calgary and Edmonton so that the next day we had another seven hours of driving or so. We arrived safe and tired. It appeared that John and Irene had done a good job of taking care of the animals and the house. Everything looked inviting, if a bit cold. Actually I think the upstairs was still comfortable, so we stoked the fires and got everything comfortable.

Now we could get back to our normal operations, but we did have one more trip to make, actually two. Kurt wanted to attend the Maranatha Bible School in Minnesota. Carol Peachey had gone there, and she encouraged Kurt to do it. I think Joy and I both also encouraged him to do it as he would then meet some other young people and have some fun. He was always inclined toward Bible studies and organized churches, and that was fine with me. So we needed to take him to Stirling, in southern Alberta. Stirling had a Hutterite colony, and some of the former members were now Mennonites, and a number of their young folks were going to Maranatha. Kurt would go with them. And on the way back, Kurt would be brought back to Baytree by some of the Stirling people, and they would just visit there as well. We made it out that on the way down we would swing around McBride and visit with some of the Mennonites in that community, particularly Joni Byler.

We just got turned around or arrived at home when we had to again go south. On January 5 we headed southwest through Dawson Creek and then south through the Pine Pass, past Prince George, and arrived in McBride toward evening. McBride is a small city, maybe 1,000 population nestled against the Canadian Rockies on the west side. It was a lumbering community as there were beautiful cedar forests in the area. That was the "forest products" that were to come out of the community. And with that activity, many of the workers were rather straight folks, not churchgoing types, so the Mennonite community was finding it a bit uncomfortable there. We stayed overnight with the Bylers in their nice comfortable quarters, nearly right at the foot of the tallest mountain in Canada, Mount Robson. I remember driving through along the approaching highway to McBride, and there were dark groves of Cedar trees along the road. Tall and with just a tinge of red on the bark. Cedar trees, being a close relative of the famous Seqouias and Coast Redwoods, could grow to astounding heights, as tall as 80 meters or more, and their lush evergreen branches and needles were like in a fairytale. It was very impressive. I also remember that Ben Schlamp, as he was moving to Baytree, stopped for six months or so in McBride to check it all out. I think he may have had some thoughts of logging in that area, but McBride was also a tourist area, and the land and housing was starting to become expensive. Ben told me he left the area because "one couldn't live on scenery." Ben ended up in Baytree.

The highway through McBride went through Jasper and east to Edmonton, so the next day, early in the morning, we headed east, through Jasper and on toward Edmonton. We then went south, and by the end of the day we were in Stirling, Alberta. Stirling was maybe 35 miles north of the Canadian border with the US. We were actually retracing our trip to Shelby on the same road, and Stirling was just off that highway. It was east of the Canadian Rockies, a bit rolling and part of the Alberta plains, not many trees and then only in river valleys. We were put up in one of the homes of the Mennonites, but nearby one could see the large dormitory-like buildings for the Hutterite community. The Hutterites were communal with no private property; all the members shared everything material. They lived in communal buildings, ate in a communal dining hall, and attended church services in a large building. That was the same as "regular" churches. I was not told why some of the people left the Hutterites and jumped to the Mennonites, and I felt that I really didn't want to know. It would probably be some dreary stories or whatever.

I think we headed home early the next day, without Kurt. I know I felt a bit lonely already just leaving him with strangers, and Joy too was hesitant. But he was now going on 18, and this would finally be good for him. He was at Maranatha a week, I think, and he came home enthused and happy. I found out that there were many young people at the school; it was very organized with Bible study sessions every day and singing every evening. Kurt was a good singer. He enjoyed the A Capella singing. He had a low singing voice, not a bass but nearly so, and it was very nice to hear him. He brought a tape home with him that had the singing of him and his new friends. We hurried to Dawson Creek and got a tape player, and that tape was in use every day. I am not sure if we made copies of it, but I hope we did. Kurt would still have that tape, I think.

He also did quite a bit of studying the Bible, and I believe he felt it all worth his time. His tuition and expenses were mostly what he had saved working for Gary Henderson, although I hope we gave him some help, what we could. And I also remember him telling us about the various preachers and bishops who participated in the program. One I remember was an Ernest Bontrager, from somewhere in Oregon. Ernest was very outspoken against drinking coffee. He considered it a poison, and I do know that he was not alone on that thought. Somehow caffeine is described as a drug, so that in itself makes it "evil," but he really made an impression on Kurt with that. And he seemed to be somewhat of a charismatic speaker. Many preachers end up getting an air and a delivery that makes them imposing and important. I wondered.

By the third week of January 1980 we had settled down to our routines. Joy was back doing housekeeping during the weekdays; the children were now going to Bonanza school, including Charity. Bonanza had a Kindergarten so Charity went for that. And Kurt was still working on his correspondence course but was often at home and also working for Gary Henderson or Jess Peachey.

CHAPTER TWENTY-TWO
Working in the Bush

In our efforts and thoughts on becoming sustainable and also to just make it from one day to the next, it seemed that logging and making lumber were a sure way to help ourselves. We were planning on putting a substantial permanent addition to the log house on the front, one that would give us more room. Since we had six children and they were all growing and wanting some space, the log home was getting very cramped. So an addition would be nice. But that would require building supplies, such as lumber. We would also need some other materials like plywood, concrete and windows. But having a nice supply of lumber would be a good start on the addition. Besides, one could always sell some lumber to a neighbor, and that was a promising enterprise.

In late January Kurt and I drove to Spirit River to register for a timber permit, like we had done the previous winter. January in the Peace River Country was mostly dark, gray skies and white ground. And cold. So we made the 80 kilometer trip over the frozen Highway 49 grade. As we approached Spirit River, maybe 10 kilometers out, we had a paved road. Actually one could tell it was paved as it was more rough than the frozen dirt grade. Not that it was all rough, but in the north country with cold weather, snow, rain, and freezing and thawing cycles, the 8-centimeters-thick or so paving just got beat up bad and heaved and cracked and occasionally just fragmented. And there were also lots of potholes to avoid. So the pavement was not the full answer to transportation. It would have been good to have a train running from Dawson Creek to Spirit River; there was still a grade constructed in the 1910s. But Canada was as bad or worse about understanding transportation solutions. The "private vehicle" is just ingrained into the culture and maybe even the DNA of some of the people, surely not the human beings, wherever they are.

So we received a local timber permit that entitled us to cut 15,000 board feet of spruce and pine in designated areas. I think we may have had to pay $150 or one cent per board foot, a ridiculously small amount. And we were instructed to not cut more than 10 percent over that, thus up to 16,500 board feet. We would have to pay a penalty of not less that $50 or more than $500 if we overcut. So it was almost an invitation to rape the land. In the US that would occur, but this program was intended for the small farmers and citizens like me. And Canada did make an effort to hold the line there. The big timber companies had their own lands marked off and had much more than enough anyway.

There were other restrictions and instructions; logging shall only be done in designated areas, no debris from loppng (cutting branches from a fallen tree) or milling shall enter any water stream, no gasoline or deleterious material shall enter any water streams, and a 20 foot debris-free perimeter shall be kept around the cutting area. So with that we were ready to cut some logs. There was a designated area for logging near the site where we were the previous year. And Jess Peachey still had his sawmill set up there. We could cut the trees and skid them to the mill conveniently.

The weather was turning cold again toward the end of January, but logging in the bush could be done at temperatures down to say -50 degrees C as long as one wore layered clothing and gloves and a ski mask to cover the head and nose. And getting started would always be the hard part of it all, just getting moving when it was cold. Once the activity started and one moved around, it would actually get where the outer layer of clothing may be shed, not always but sometimes, particularly if it became sunny and warmed up to say -30 degrees C.

And about this time Ian Marchuk contacted me, came to our house one evening and proposed that I could go with him to cut a seismic line across a ravine north and east of us. He had a contract with one of the oil or exploratory companies to, by hand, clear a line through a ravine or creek valley. He told me that they paid by the job and that I would get $95 for the day's work. That sounded OK to me, and we planned on going out a few days later.

So Ian came bouncing up our lane early one morning; it was very cold and dark too. I don't remember what time it was but probably about 6 a.m. or so. If we were to cut a mile-long path, it would take a good amount of time, and we needed to get started early. I don't remember if Joy went to do her housekeeping but probably did. So I bundled up, with plenty of clothing and a good ski mask and another pullover cap. I sure didn't want to get cold. And I had my chainsaw, gas can and some lunch. Can't remember what I took along to eat but probably two peanut butter sandwiches and also a jug of water. Ian was driving his pickup truck, and it was like many of the vehicles up in the Peace River Country: The windshield was cracked, he only had one headlight and the other one was broken, and there were dings on the body, all from flying rocks and from driving on gravel and dirt roads, passing or being passed by big trucks and speeding pickups. The country was made for destroying a vehicle, which was a shame too, because in Ohio and

Indiana the roads were paved and flying rocks were a rarity, but the salted roads in the winter started the vehicles to rust away, and the summer rains and hot humid weather just sped up the process. The vehicles would just get eaten away, while here in Canada, if you could keep your car or pickup from getting hit by flying rocks, the roads were not salted, and it was essentially a dry country. Rusting was not common, but dinged-up vehicles were.

We must have driven for an hour in the dark going north up a dirt road, but all at once it just ended, dead end. Ian stopped and I think kind of parked along the side, although no one would have come this way, and they too would have had to stop. So we got out of the truck and took our saws, gas, lunches, water and a small toolbox. There was a bit of a trail going down, north from the road's end. It was really the Bush: many poplar trees, and some birch, willows, and the odd pine or spruce. It was a typical Peace River Country ravine. The road had approached the ravine on flat, level land. And right there it was a sudden descent to the bottom, some 800 meters down to a creek or small river that we would not see for a couple of hours as now we were just trying to get down the incline to a point where we could begin clearing. Well, we went maybe 80 meters down, me carrying my saw and lunch and Ian with his saw, lunch and toolbox. We were walking side by side and as we turned, to me he appeared different. I asked him if he had some sort of nose protector on; his nose was white as alabaster or meershaum. I had just thought he had a nose protector. And I can still see his face. "Oh, no." I hadn't noticed that for some reason he did not have his ski mask on, just a pullover cap. And as we descended from the truck at maybe -50 C, we were going rapidly down and had dropped maybe 30 meters in elevation but to -60 C, maybe. He had frostbit his nose. Well, actually the weather had frostbit his nose, but it was all the same. He quickly covered his nose with his hand and rubbed his nose until it got some color back. We decided right then that we needed to get a fire going to get warmer. As I said before, getting started was the hardest part of working outside up here. And we had not really gotten started. So Ian and I gathered what sticks and brush we could find, and that was not easy. The snow was maybe 30 centimeters deep, and we couldn't really tell what was good firewood and what was bad, green or wet. But we kept moving, and in 15 minutes or so we had a reasonable pile of sticks together. Ian dumped some gas, gas mixed with oil for his two-cycle chainsaw, on the pile and lit a match. The sticks did start up, but it was not a pretty fire, one you would stand around watching. It was mostly just gasoline burning on some sad sticks that had just been there by mistake. But the flames did provide a bit of warmth, and we stood there, 80 meters down the hill from the truck, just trying to keep warm.

We were going to get paid (I hoped) for clearing a path from the top, where Ian's truck was parked, down to the bottom of the ravine and up the far slope to the top. There was supposed to be a road dead ending there. If one could imagine, a long 1.5 kilometer bridge straight across would then connect the two dead-end roads, progress maybe or better yet, another tower of Babel. But there was no bridge yet, and our task was to make the surveying crew's path clear for seismic line studies. That didn't impress me, but it was a beautiful ravine, and with all the trees draped in snow and the land covered with a blanket of white, that impressed me. The work started with just using the chainsaws and cutting brush and small bushes away, mostly just clearing so someone could walk our path without too much effort. The work was a bit tedious, and the biting cold became bearable, and we even started to get warm. Ian's nose had recovered nicely as he displayed around noon when he took his ski mask off. The nose is a delicate thing. If there is enough hot air, expiration moving around it, the nose can endure cold well, but still cold air and also some negative radiation leaves it to suffer. Ian found that out, and it doesn't take but a very short time. In engineering we use a term called the time constant which is a time before a glob (like a nose) reaches the temperature of its surroundings. I think the nose has a time constant of about 5 minutes left to itself. Wear a ski mask, I always thought.

By noon we had reached a point near the bottom. The descent was not uniform, but rather there were some almost-level sections and then a sharp drop off. We even encountered a few large boulders in the path, not too common in this country. We went back up to the fire, which had mostly died down to embers and crinkled twigs, and had some lunch. The temperatures were probably in the order of -20 C, and it felt warm. The sun had come out reluctantly, and we even sat down and discussed some sort of politics, social issues, or maybe just how to get by in the country. Ian was single but was trying, like us, to get some sort of a vector or direction for a long-term sustainable financial situation.

After the short noontime break, we completed the task of clearing the line. The creek or small river at the bottom was frozen over completely, so it must have been rather small. The incline up the far side was less tedious or difficult, and by 3 p.m. or so we had reached the top. As we retraced our path, we did a bit more grubbing out, and when we crossed the creek at the bottom, at maybe 4 p.m., it was getting overcast and dark. The sun would officially be down shortly, and with overcast skies, that meant darkness would descend upon us in short order.

Scrambling up the original incline, we gathered the toolbox and other supplies, returning to the truck as darkness had settled in. And sure enough, Ian's truck had two flat tires. What a coincidence. Ian said he knew the family up the road, which I had not even observed. So he took off, and I stayed with the truck. It was getting cold again, and waiting in the cold is even colder. This would test my layer preparation, and as it turned out I remained reasonably warm. After about a half hour I saw lights approaching, and the pickup pulled to a stop next to Ian's truck. I was cold enough that I didn't remember the name of the man who drove the truck, but with Ian's spare tire and another tire provided by the neighbor, we had Ian's truck back in business. It was probably 6 p.m. when we left, and with another hour of travel to our home, it was a long day and all outside in cold weather. But Ian assured me that he would get me my check for the day's work as soon as he got paid for the subcontract. And it was nice to get home to a warm room, a nice supper prepared by Joy and the girls and just getting the various layers of clothes off. I felt good about the day's work.

Then, within the next week, Kurt and I headed to the designated logging site to begin cutting our timber. I think maybe we had a couple of log chains with us with the chainsaw, so one of us would cut a tree down and the other one would wrap a chain around the butt of the tree, near where we had cut it from the ground, maybe hook another chain to that one so we had a nice long chain. Jess Peachey and maybe one of his sons, or even another neighbor, was there with one of their tractors to pull the logs to a pile toward the clearing where the sawmill was located. The timber, all spruce, was nice and straight and not heavily limbed. Actually much of the trunk was clear for the first 6 to 8 meters up. I figured the average trunk diameters were half a meter, and if we were to get 15,000 board feet, we would need in the order of 45 to 50 trees. The first day I think we probably cut half of that amount, around 35 trees. And all the time I just felt sad about my activity. It did, however, make me think about the past settlers and people who had to scratch for a living. It seems to be part of us all, finally, that anything goes if you perceive that your very existence is at stake. This is all part of our self deception, however. We can survive on so much less than we want that it isn't even a comparison. And we justify our behaviors in all the ways that are finally self deceptive; it will all run out finally, and we will become the extinct species. The evidence is becoming more and more overwhelming that we are indeed exhausting our inheritance, and the meek will inherit the earth.

Ian came around one of these evenings with my paycheck for the seismic line clearing. It came to $94 and that was a good piece of news for me. But Ian had bad news. It seems that the day we were doing the clearing, someone broke into his house and stole a number of his items. We wondered if the two flat tires were part of that. And he was convinced that it was a neighbor with sticky fingers who knew what he had and also that he would not be home that day. Thinking back, I know that Ian had made the connection for doing this through Ralph Tompkins who seemed to know all the oil people. So I had some ideas, particularly as things developed in the remaining months of our lives in Canada. But proving a theft is almost impossible, and it's not even worth dwelling on the scenario. Ian never did know who had done the work.

We probably rested for a day or so, as it was cold and the logging was hard, intense work. The next day brought warmer weather, and while we had nice hard surfaces to skid the logs over, there was a good chance the bare ground would take on the viscous tenacious characteristic that could cause problems in skidding. So we drove the John Deere D to the site, intending to use it to skid logs down to the sawmill. John Trinier and Doug Scott, one of the local neighbors, came out to help as well. Our cut logs were piled up about 800 meters from the sawmill, and we planned to do milling soon. The site was a good ways away from our home, maybe 15 kilometers, and the old John Deere was the slowest of our tractors. It could probably do 10 miles per hour, 16 kmph, and that meant a good hour's drive. But the tractor had many redeeming features. It was ponderous, moving slowly but surely. The engine was 501 cubic inches, and for a modern engine that would translate into maybe 400 to 500 horsepower--another delusion, because the modern tractors are relatively high speed and low torque. And the 501 cubic inches were with two cylinders, big cylinders. And also with that much mass moving, the engine idled at maybe 100 revolutions per minute and had a maximum speed of about 500 revolutions per minute. So it was slow, slow, slow. But it would run forever if need be, and it could pull anything at slow speed, somewhat like my two Belgians, which gave me two horsepower honestly and then some. The John Deere D was rated at about 30 horsepower, and that was at least approaching honesty. It still could not out pull my team of horses for a short burst but could go all day. Bess and Morning Star would need a rest, just about the time I would need a rest. But the "modern" methods don't want rest, and the newer tractors even have lights so one can work around the clock, well, not work but ride around the clock. It is difficult to equate riding on a machine with honest work.

That day we completed the cutting down of an amount of trees that I thought would provide the 15,000 board feet of lumber we signed up for. And it went fast with two extra pair of hands to finish it off. It was a task that I felt was like butchering a hog or a steer. And then the final cutting a dead animal up into little pieces took on a mechanical, inhuman (or maybe all too human!) character. Once a tree was felled, it no longer was a live sentient being and was only a mechanical material. Maybe that is what we too will become, but who knows. And the day had become warmer, and the skid path was viscous and tenacious. Jess Peachey's tractor was not negotiating the skid path well at all, but I remember the John Deere just slowly pulling the logs to the sawmill. One pull in particular was noteworthy; there were a few, maybe five or six, logs stacked together, and we chained them in a bundle. But they were in a ditch, and that ditch was especially tenacious mud. The John Deere, getting down into the mud, just slowly pulled the logs out and did the task without a slip or spin. Yes, it was slow, but it did get done. I think the D earned some warranted respect that day. Before then I know that most of the people would have thought me a fool for even getting the clunker. But they were not mechanical engineers, and how could they know, unless they sometime had actually worked with a slow plodding John Deere D? I felt good for the tractor, even though I knew that my two horses could have done the same work. I didn't bring them that day because I felt it was too far for them to walk and then back again. But I would get them too working on hauling hay.

CHAPTER TWENTY-THREE
Hauling Hay

The horses were doing good in the cold weather. Even the young stud, Prince, seemed to be not having any problems, and they enjoyed playing in the snow, running among the poplar trees and then standing and getting sunshine, when the sun did shine. But I suspect they were missing doing some substantial work too. So in late January and early February I made it a point to work them every day that I could. I would hitch them to the green wagon and go down the road to George Blakus's and get a load of hay. Usually around 50 bales filled the wagon up since I had left the side boards on which prevented stacking the hay too high or wide. After loading the hay I would take them home, and I began staking the hay at the straw pile and next to the granary. That was next to the Twelve Mile Road and would be convenient for anyone wanting to purchase some hay.

During this time Joy was working full-time on housekeeping at Tompkins's, and the other children were at the Bonanza school. Kurt sometimes was home but most often was working with Gary Henderson or one other of the neighbors. He was building up a reputation as a good reliable worker and was in demand for just about any work to be done. So it was mostly me doing the work with the horses, and that was OK. I somehow never was one to be lost when I was by myself. And the horses were good company as well as Prissy, our dog. She too would go with me when I made the short journeys to haul hay. And I also included Prince in the journey as he would feel left out otherwise.

One day in early March, I think, I was at Pilsworth's just visiting and Doreen Higgs happened to be there. Fred was probably playing a game of cribbage with Doreen and drinking a cup of coffee, a popular local beverage. It was a good break to just talk with some people as I think I had gone maybe three or four days by myself, except when everyone came home in the afternoon. And the conversation somehow took a turn to horses. Fred was a horse admirer, and Doreen too felt a comradeship with horses. She apparently had grown up with riding horses. So whether I suggested it, Fred suggested it, or Doreen did, she offered to help me harness the horses and handle them for the hay hauling. It all seemed fine with me, so we set it up that Doreen would come over around 10 a.m. to help me work the horses for a couple of hours.

We settled into a routine of Doreen coming over, and we would go out and harness the two horses, hitch them to the green wagon, hook up Prince next to Bess, his mother, and off we would go to get a load of hay. And every day became more comfortable for all of us. This was one of the things I had envisioned long before we moved to Canada: working horses regularly. And I found out that the horses just became a part of the effort; they were real team members, and with their intelligence, any problem could be avoided. They got so that they anticipated a potential bank of snow hiding a deep ditch, something I would not know. And how they learned it or whatever was and is beyond my simple mind. Most people would say it would be instinct and then go on praying to their dead god.

And after a good workout of two hours or so the horses were ready for some oats and fresh hay and being tied in the barn stalls. They would be tired. And so would we. I would stoke the kitchen stove and even the pot belly to get the downstairs nice and warm, and Doreen and I would sit and talk about the day's activity, kind of debriefing each other. She was a good horse person, and I learned much from her, much of which I would probably have gotten alone but it would have taken longer. Now I would have let Prissy in the house, but back then Joy was adamant about dogs being in the house. She would not have it, and Prissy and I understood.

I suspect that because Doreen's husband Reggie was a truck driver and gone for long stretches of time, she too appreciated company. She and I were separated by probably 15 or 20 years, and there were never any digressions; all at a very platonic level which made it simple and comfortable too.

Even with a telephone for contacting people or being contacted by people, it was not used too much. I had the address and name of a contact person at Montana Tech where they were looking for a faculty member to teach mechanical engineering. I was just going with the flow now, after visiting in Indiana and returning for a visit to the University of Dayton. It would be a disappointment to give up what we had worked so hard for during the past 18 months or so, but reality was beginning to nudge into the picture. So I wrote a letter to Montana Tech and sent them a recent resume, explaining in a cover letter what I had done during the past 18 months or so. Academia and businesses too are always looking to find out why someone would not be presently in a position that they would like to replace. I provided enough information so they could call me or send me a letter. Whatever. I was really ambivalent at that time.

Also, I had an address and a contact person at the University of Wisconsin-Platteville who too was looking for a faculty member to teach mechanical engineering. And I sent that information out as well, just to see what in the world would transpire.

And then, for some reason I was contacted by a local logger, Ross Chilton, who wanted to build a sawmill in the bush south of Grande Prairie. And a few days later a person called from Spirit River and wanted someone to design a heating and ventilating system for an apartment complex there. I had made up a business card and placed it around the community to see if I could drum up some consulting work. It seemed to have worked.

As it turned out, the apartment complex in Spirit River was just a dream and not even in the planning stages. So I could write that off as a viable project. But Ross Chilton did contact me again, and he wanted me to design a sawmill to be located about 20 kilometers south of Grande Prairie. So one cloudy, cold day (there were a lot of them!) he and I headed to Grande Prairie in his pickup truck. Grande Prairie was just beginning to become a real boom town. It was sitting in a collection point for all the potential "new" oil that Canada seemed to want to spend. But at this time it was a town of perhaps 20,000 people, maybe twice the size of Dawson Creek, but it had the same feel as Dawson Creek. As we headed south out of Grande Prairie, the number of roads became fewer and few, until at about 10 kilometers south of Grande Prairie there was only one north-south road, a two-lane one that turned into a gravel or graded road after another five kilometers. We turned west on a single lane dirt road, and after another five kilometers or so we reached a crude clearing among the typical Boreal forests of the area. These were destined to become an endangered community, like all the regions that had any potential for oil. But at this time they were still beautiful mixed forests, poplar, birch, pine and spruce. And this was where Ross apparently had financial investments or interests in building a full blown "forest products" operation. We surveyed the area, and he gave me an idea of the size of operation he intended to reach. It was ambitious, although finally it would be, in his mind, a typical sawmill operation with a large conveyor/table for handling up to maybe 15 logs, each log to then be placed on a large carriage, one at a time, and this carriage would travel back and forth feeding the log into a large circular saw, the sawmill. The log would, on the first cut, have a rough slab cut lengthwise from it; returning, the log would be rotated 90 degrees so that flat surface was now on the carriage bed, and the log would again be carried into the saw blade. This time another slab would be removed, and again the log would be carried into the saw. This time a board would be cut off, maybe 5 or 10 centimeters thick. There would be one rough edge to this board, but that would be trimmed off on an edger, which he thought I could design as well. Progressively, boards would be cut off the log until there would be only about half of the original log left to saw. Here the log would be rotated another 90 degrees so the flat surface would be placed on the carriage bed. The log would then be completely cut into boards in this position. Nothing new here, but it would be interesting work to get the plans worked up for his construction. As we talked, I could tell that he was not really a mechanic, a welder, a machinist or anything other than a person with a mouth and apparently a pocketful of money. I remember him referring to the amount of I-Ron needed to build the mill, and I thought to myself that he had no comprehension of the amount of work involved in coming up with a design, on paper, so that machinists, welders and mechanics could put it all together. My business cards must have impressed him.

On the way back to Dawson Creek he indicated that he wanted me to begin right then and get his sawmill designed. He seemed to have known that he needed some "drawings" to get his mill, and that was what he wanted me to do. And he started talking about religion. Well, Numerology to be precise. He seemed to know a bit about it and listed some names that were "strong"; Chilton and Ross were among the strong names. Other names seemed to be moderate or weak. My name was moderate, or at least that was what he said, maybe just patronizing me. I had no interest in all that as it seemed to be the worst part of Astrology, and I never was intrigued by astrology. So if he was trying to convert me to his thinking, he was wasting his time.

I had brought my drafting equipment with me, including a tee-square and a drawing board. It didn't take much space, and it was one of those things I thought may be useful. So we returned to his home in Dawson Creek; he had an area that would work for doing the work, although it would be more time than I think he expected. I told him his plans would not be forthcoming in less than a week, and that would be pushing it. So I spent three hours beginning a layout of the footprint of the conveyors, the carriage, the sawmill and arbor, and the gathering table on the end of the mill. I think I had made a good bit of advancement toward a design. He actually left for an hour or so, and I was by myself in his house. He seemed to be a trusting type, or whatever. When he returned, it was time that I got home, so he took me back to Pilsworth Corner and our home.

I returned the next day to spend another three hours or so on the design. He had plenty of drafting paper, and I filled up maybe three sheets. I had a layout of the placement and relationship of the components and another two elevation views of the mill itself, one an end view the other a side view. Ross didn't seem to know much about reading drawings, and at that point I told him I needed an advancement, a payment to cover my costs. I told him I would send him a bill. I didn't hear from him again, and for some reason I was not surprised....

I had my horses to work, and taking two days off was more than I should have. Bess and Morning Star were still in good mentality, ready to work. So Dorene again came over and helped harness them up, hitch them to the wagon and drive them up Twelve Mile Road to the hay pile. This was work that I could do, and I didn't need to have a numerologist trying to cheat me out of more time. Our pile of hay by the granary was now getting large, and I was hoping someone would need some hay. I posted a sale ad at the Coop in Dawson Creek, a place I thought would be where potential customers would see it. Not a nibble, even though we now had a telephone. So I was just biding my time.

Then in the second week of March I received a telephone call from the University of Wisconsin-Platteville. They did not need me now; they already had a new faculty member committed, so I was now down to Montana Tech. That was better news as they called about a week later and wanted me to come to Butte, Montana, to the Montana Tech campus for an interview. It seemed that the earliest we could get together was going to be the last of April or the first part of May, near the end of their semester. That was OK with me, and I think we made out that the last week of April would be good. I had lots of other projects going, but I was relieved to know I had some viable options now. Even though Montana was a long way from Baytree.

And we continued to visit with neighbors, the Blakuses, Fred and Irene at their home, Werner, Jess and Faye Peachey, Willard and Arla, or the Sadliers. And Joy was working full-time now at Ralph Tompkins as his housekeeper and meeting all the oil men Ralph associated with. We started getting to hear the various scuttlebutt, rumors and whatnot. Not good activities. Joy was getting suspicious of me and Dorene working with our horses. I have no firsthand knowledge of this, but Joy did ask me if there was anything going on between us and that whenever two people, a man and a woman are together, and together alone, things can easily happen. Such thinking and gossip were not helping our relationship at all, Joy's and mine. Since I had enough work and projects to do, I was able to submerge any of my negative reactions to her statements, and Doreen continued to help me with the horses.

Another bit of information I found disconcerting was that apparently Proctor and Gamble, a large corporation with headquarters in Cincinnati, Ohio, owned a very large tract of the boreal forest south of us in the Saddle Hills. So here we were in northern Canada, and a grubby, greedy US corporation held ownership to some of the best parts of Canada, right across the road from us.

As March became longer, spring came up on us, only on the calendar, however. March 21 was still cold, but there was a hint of warmer weather coming. Chinooks were always a possibility, and they made the long winters more endurable. But having a long stretch of warm summer weather would be welcomed. One of the natives, maybe George Blakus, told the joke "Canada only has two seasons, winter and poor sledding." Accurate, but I had become acclimated and had no problems with the winters, the cold or the snow. Yes, summer would be welcome, and it would be poor sledding. It would be something to tell everyone, so only the hardy and adventurous would come to the Peace River Country. It didn't need the beautiful people, and I didn't either.

Jess Peachey was ready to saw some logs as he wanted to get that done before spring planting and the warm weather activities. So sometime in the latter part of March we worked for a few days at Jess's mill. Kurt and I spent maybe three days there, and I think John Trinier and some others came as well. Sawing lumber was best done with lots of hands. Working the woods, cutting trees, was best for one man. In the pre-chainsaw days, a two-man saw required two people for cutting trees. And that was better because finally having two people watching a tree fall is better than one. But with the chainsaw mentality a tree could go over, and if two people are there one may not know what the tree is starting to do. Even one person can get into trouble. But with the milling, many hands made light work. Jess was the sawyer, the mill operator, and he knew his skill well. He could get the first slab off straight, rotate the log 90 degrees handily, and cut 1 inch, 2 inch, 3 inch or whatever thickness board was desired. And getting a uniform thickness of boards was an art. He did it well. But the others made his work smoother. There were always boards coming out of the mill, and they needed to be stacked or at least moved out of the way, a fresh log needed to be ready to be positioned on the carriage, and wood chips from the cutting saw were continuously piling up under the saw blade. They needed to be scraped away. And there was always the task of just keeping track of whose logs are being cut and what size to cut them, and how many of them. So four or five people could be busy all day, in a community effort. After this I remember tallying up our boards: (all inch units here) 1 x 4s, 2 x 4s, 1 x 6s, 2 x 6s, 1 x 8s, 2 x 8s, 1 x 10s, 2 x 10s, 1 x 12s, 2 x 12s, 4 x 4s, 4 x 6s and some 8 x 8 timbers. I would have spent over $15,000 on all this in a lumber yard. We tallied up 16,212 board feet of lumber, all spruce with maybe a bit of pine included. We would plane the boards to make them smooth later in the summer. But for now there was the need to pile the boards not tight but stacked, staggered so air could get between the boards and dry them all out uniformly, in principle. We hauled the boards to our farm and piled them down near the hay pile, the straw pile and the granary. This was satisfying work and productive too.

Sometime during late March or early April, it was decided to get a television set. Ralph Tompkins had a television set, and so Joy as able to watch there. And the children at the Bonanza school were watching television at home. Most all the neighbors had television, except for the Mennonites. Even then there were some of them who "unofficially" seemed to have a set somewhere. So we got a small set, one that would not take too much power, and we bought an extra car battery with a heavy cable set and installed the set in the hallway at the top of the stairs. I think the battery, when fully charged, would run the set for maybe four hours or so. At that time we needed to put the battery in a warmed-up vehicle, the pickup or the one-ton, and change with the one in the vehicle. After maybe two hours of driving, the battery would be back to where it could run the set.

So we caught up on our culture, or whatever. I never was a fan of television, and I would not have gotten the set. But there does come a time when others need to have their say, and this seemed to be one of those times. I remember some of the neighbors saying it was good for the children, although they never explained what was good for them. I do admit that there were some enticing programs that I too would watch with the family: Colombo and The Waltons were entertaining, if not enlightening.

The only thing we now lacked was electricity, and that seemed to be more because of the expense of connecting up as well as the fact there was a backlog and we would have had to wait probably six months to get it. We never did seriously discuss getting that amenity, again, if it is an amenity.

CHAPTER TWENTY-FOUR
A Delay

During early April I was planning and anticipating the spring and summer work. Firewood was always needed, and whenever I had the time, that was the next priority. And it was getting to where I was advancing, or at least it seemed like I was advancing. My horses were still working with me to get the hay all hauled to our property. There was a bit at Pilsworth's, but the hay at Blakus's and that up the road across from Marvin Meyers's was all on our property. So I had a bit of a breather and decided to make a more substantial fence for the horses, one that would let them graze down near the granary. I actually planted some posts along the Twelve Mile Road and then could string barbed wire there too. I tried to use willow for posts as they seemed to be the best. Spruce or pine would be poor material as they would rot away even up here. Poplar was tempting, but it too would not hold up too well. Willow, on the other hand, was mostly small and gnarled and crooked, not attractive fence post material. But I managed to get enough willow and poplar to run a decent fence from the barn straight north to the cleared land, then east to near the road. I was not sure how I wanted to fence in the cleared land, but I went ahead and fenced south again to our lane, and that made an enclosed grazing plot for the horses.

And I was on good speaking terms with the horses. We had been working together now for 20 months, and the past three months were intense, almost daily workouts. They were comfortable with me and I with them, unfortunately. While I was building and repairing the fencing, they were mostly right there inspecting and checking it out, all three of them, Bess, Morning Star and Prince.

One day, it was getting toward late afternoon in early April with still snow on the ground and cold air, and I was just finishing up a top row of barbed wire for the fence near the northeast corner. It was where a patch of willow bushes were clumped, and I started back to the house. I think I may have just left the fence stretcher and a bucket of staples and a hammer at the corner by the granary so that after a break, I would come back and clean everything up, checking to make sure the fence was substantial. The three horses were with me, and we started back to the house and barn. They knew where we were going, and I think they were probably happy to be going back to the warm cozy barn for a cup of oats, maybe, and some fresh hay. We now had plenty of fresh, good quality hay. So Bess and Morning Star were side by side headed back. I could tell they were happy and letting off some steam that had accumulated since we didn't do any hard work that day. Morning Star was on the left headed home, and Bess on the right. And they just got to kicking each other, rather playfully, and having some fun too. This is all that I was taking in but not thinking about very well, if at all, because I walked up next to Morning Star on her left and told them to go on, let's go home. I no more than said it and I found myself flying through the air, literally maybe a meter off the ground headed east, directly away to the left of Morning Star. I was not, at that instant, aware of what was happening. The only thing I remember, and I still do, was seeing a poplar trunk, maybe 20 centimeters in diameter coming at me. It seemed to be slow motion, and I can still see it clearly. I knew I needed to grab the trunk of the tree to stop myself from going any farther away from the horses. I did. I grabbed the tree trunk with both hands wrapped around it, and I rotated one and a half complete turns around the tree. I slumped to the ground, still holding onto the tree trunk. My momentum had given way, and I was still. This all happened in an instant, and then I knew what had happened; Morning Star had kicked me good in my right side with her left hind foot, just playfully. But I was not Bess who could take a playful kick. And I sure could not kick her back. I could not move. My whole body felt paralyzed and in pain. I was having a hard time to just get my breath. And then it was only the constant pain, mostly now in the right rib cage. I laid there. Then I realized that I could call out. I think I probably yelled or tried to yell, "Help, help, help." It was not long until Kurt was there above me. Faithful Kurt was always alert to things, and I think he saw the three horses coming to the barn and ready to be fed. But I was not there, and he put it all together and came after me. How he knew to go where I was, I don't know. And I think maybe it was 15 minutes or even half an hour before he came after me, but I had lost track of time. I was just trying to get my breath. Even the cold ground was not important or significant at this time. He saw that I was still alive, and I think he covered me with his coat, even though I had my coat on. Lying on cold ground and not moving quickly becomes life-threatening cold. And he had the presence of mind to cover me better. He told me he would be back quickly with the truck. I was not going anywhere by myself now.

Shortly, at least that is how it all now seems, Kurt was back with Joy, and they had the pickup truck there. I think maybe one of the neighbors was there as well. Joy probably called the Pilsworths or Blakuses, and one of them came over as well. They all seemed to have had the full understanding that this was serious and everyone needed to be careful, yet I had to have professional help too. They kept asking, Joy and Kurt and maybe George, whether I could move anything, could I breathe, could I move my neck, and so on. I understood all they were concerned about, and so was I. And I knew that outside of my right lower rib cage, I was now OK. I just could not get a good breath of air because of the pain associated with that. So, somehow they had a plywood sheet, 4 feet by 8 feet, and they carefully slid it under me. I was nearly immobile but rolled over, probably on my left side, and they slid the sheet under me from my right side. Once I was on the sheet I think Joy maybe put a pillow under my head, and the cold ground was now gone. Even a half-inch-thick sheet of plywood provides some insulation, if temporary, to the cold, and so they then lifted me into the bed of the pickup truck, closed the tailgate and we headed to Dawson Creek. I am sure Joy had called ahead, maybe just before we headed out, as she wouldn't have known what had happened before she came down. Kurt told her to hurry down with the truck, so she probably called the hospital as we left. The ride was cold, bumpy and uncomfortable, but we made it. And I remember that I was still just trying to get a good breath.

At the hospital we went to the emergency room, and in their rigid, formal, efficient manner assumed that I would need some X-rays. A doctor came in and looked at me, talked with me to ascertain the reason for my condition and after I explained the happening to him, he said "No need to do an X-ray. All of his ribs on the right side are broken, and the only thing we can do is help him to heal." I was not able to move so I could see it, but Joy told me that there was a circular redness on my right side about 20 centimeters in diameter, the right size for Morning Star's hoof. It was right between the first seven ribs, just low enough that the top or first rib was probably just badly bruised. The others were all cracked or broken, but, as the doctor observed, there was really nothing to do except try to get as comfortable as possible and heal, without contracting something more serious. I suspected that even the eighth, ninth and tenth ribs were bruised. And everything just hurt. But I was probably very lucky because if Morning Star had really wanted to move me over, she would have hit me harder, and that would have done me in. As it was her playful kick was not fatal, but if she had somehow kicked me higher up, say the first four ribs and above them, I probably would not have been in an emergency ward either. I still had my lungs intact and not collapsed because that is one of the purposes of ribs, to protect the lungs from being collapsed. At the time, though, I felt like my lungs were collapsed.

They carted me to a ward, probably on a four-wheel stretcher, in a room with a double bed, and I had the inside bed, next to the hall. I was to stay for three days in this room, just trying to catch my breath and not contract pneumonia or whatever else. Hospitals are notorious for being full of disease and germs, but all I was interested in was finding a position that would be tolerable. Lying on my back was not good, nor on either side. Lying on my back prevented me from taking any sort of deep, even shallow, breath. I tried lying on my stomach, and when I put my pillow on my stomach that was good for a few minutes. But everything was just surrounded by this pain in my right side. The first night was the worst as I think I did not get any sleep, just kind of drifting off for a few minutes because I was tired too. The night nurse was very helpful and expressed her concern for me. She suggested I raise the bed up so I was sitting on the bed and take the pillow and squeeze it around my rib cage. That was a position that was at least promising. I was still in pain, but I could actually get my breath, even if it was shallow. And anyone who is in such a predicament knows that shallow breathing finally leads one to want to cough, clear the lungs and throat. That was the worst part, because trying to cough and finally coughing made the right ribs all yell for calm. But I was yelling for peace.

The nurse, I think, spent more time with me than any of the others on her watch. There was another person in the bed next to the window in my room, but a curtain separated us and I never met him or her. I hope I didn't disturb the patient too much. I was able to drink water and eat some snacks, but I think maybe that first night I didn't get out of the bed. It was difficult just staying in one position, and any extra movement would have been pain without end. So somehow I made it through the first night.

The next day Joy and the children came and visited me, and that was such a welcome sight. I was brought up to date on the conditions in the farm. I had missed the milking of the previous evening, but Kurt or Tim handled that and also took care of the horses. I emphasized that Morning Star was not a malicious horse and was just including me in her play. I was just a weak person, unable to take a little bumping. And I was now able to sit up for longer stretches, and I even got out of the bed once or twice. The nurse helped me negotiate getting to the stool and to wash up some. And I did feel a bit grimy and even sweaty, but a bath would have to wait until the third day.

Toward evening of the second day the night nurse came and asked if I would like to try and sit in one of the chairs in the room. That sounded like a good idea, and it turned out to be just that: a good idea. She and I sat and talked for a long while. She had other patients to watch and check on from time to time, but after a half hour or so she returned and we would continue our conversation. I don't remember everything we talked about, but it would have included where she came from, why she did nursing, and whether she felt good about living in Dawson Creek. I think she was born and lived in the Peace River Country her whole life and was reasonably content with it all. She wanted to help people and earn her living. And she was interested in why I was up there, what I had done before coming up, and my family would have also been a part of the conversation. But I did feel an attraction for her and her me, I think. She indicated in many little ways that she wanted to get to know me more and perhaps we could somehow become good friends. Actually even after two evenings one can be a good friend, but she was a good nurse to me and also someone to talk with without being cautious. That was how I felt around Joy for a long time, even before moving to Canada; conversations with her would have to be measured and planned so nothing too different or confrontational would enter in. It didn't have to be that way, but it seemed to be a continuing part of our relationship, and communication was breaking down.

Toward morning the night nurse took me to the bathroom in a wheelchair, disrobed me, and proceeded to give me a bath. I had to remain seated in the chair, but she was very thorough in her getting me clean, and I at first felt a bit uncomfortable, but then with the constant reminder of my broken ribs, it was fortunate for me that I was not able to move with any major intentions. It was a good bath and probably one of the best I ever had. She even gave me a nice rubbing when she dried me off, and it was very comforting and therapeutic, maybe too much so. She then took me back to my room and left to look after some of the other patients. Before she left she gave me a hug and I her. She wished me well and left. I was to see her from our truck window walking in Dawson Creek one day. But I was with Joy and some of the children, so I couldn't even wave, but our eyes did meet warmly.

The third day of my visit to the Dawson Creek hospital was my last day there. I would have to go home and just try to heal and recuperate there. The doctors and nurses at the Dawson Creek hospital were professional, caring and up to date on what helps patients. I at no time felt that they were wondering if I could pay the bill, whether I was taking the pills or keeping up with any other practices or methods that they had suggested. It was all low-key, comfortable medical practice. I had and would have medical experiences, including doctors, nurses and hospital visits in the US, and the only thing better stateside was the high costs they expected you to pay or cover somehow. People who think the medical industry in the US is superior or even equal to Canada's are badly mistaken. A benign form of national socialism exists in the US for medical, automotive, aircraft and general manufacturing, not quite total but enough to make 1 percent of the citizenry having over 40 percent of the wealth and the top 2 percent have over 55 percent of the wealth. A very subtle and in a way, innovative and a bit of genius there. Not quite like the Czars and Kings and Queens, but enough that the common citizens seem to be happy and distracted. A kind of opiate for the poor and lower middle classes. Canada seemed to be at least trying to be fair. They subscribed to the principle that everyone should be entitled to health care regardless of economic situation, everyone should be paying and participating in their fair share of the national and provincial loads, and everyone should be treated as a law-abiding citizen. Three premises lacking in the US although touted as part of the national conscience.

I bid farewell to the nurses and other staff members and took the ride back home. It would be good to be at home with the family even though that first week back was mostly just sitting and watching others work and move about. I think I progressively took on activities. First I would put some wood in the stoves when they needed it. That was a chore I could handle, slowly. And then I probably did some minor floor sweeping and maybe even washing a few dishes. By the end of that week I think I was moving around outside and even going to the barn to check on the horses and Flossy and of course pet Prissy. Prissy somehow was always there and just somehow overlooked too. But we did love her.

My broken ribs were healing nicely, although it would be six months before I would be back completely to normal where I could take a deep breath without getting the reminder of my ribs' insult. And I was fortunate too in that I had no complications such as congestions or at worst, pneumonia. I had a good family and wife who gave me tender care and showed their concern for me.

CHAPTER TWENTY-FIVE
Going to Montana

After my delay in going to the hospital for the rib damages, I contacted Montana Tech and asked to put off my planned visit for a couple of weeks. I explained the situation, that I had been kicked by one of my draft horses, not a riding horse, and was getting back to normal but felt a two-week delay would be better. They were understanding, and so the second week of May I would go to Montana. And about this time it seems that the exchanging of truck batteries to keep our television operating was taking a toll on the pickup truck. It earlier was showing signs of needing a tune-up or something, and now it was getting even more aggravating. It didn't want to start consistently and seemed to not want to run reliably. One day we even took the John Deere B and towed the truck down Twelve Mile Road trying to get it started. Finally it sputtered and took off, but it was still sick. We needed to do something about it. Kurt suggested that we just needed to give it a good tune-up and that would take care of it. Joy somehow must have talked with Ralph Tompkins about our truck problems, and he suggested bringing the truck over to his garage and doing the work on it there. His garage was heated, had a concrete floor, and he had lots of tools too. I didn't like it at all. Joy was giving all sorts of odd signals. Even when we had gone to Indiana and Ohio for a visit, I somehow noticed that she had a birthday present for him and sent him a birthday card for his birthday. Not big things, but it sure made me wonder. So anyway, we decided to go ahead and take the offer. So we spent a few days working on the pickup truck. And that turned out to be more than just a tune-up. Kurt was a good mechanic and suggested we take the head off, look at the valves and the piston rings, and put in a new head gasket. We did that and as a final check, to make sure that we were getting fuel, we checked the fuel pump. Everything seemed to be OK. For some reason, we also decided to take the fuel tank out and check it. This model of truck had the fuel tank located just behind the bench seat of the cab, inside the cab. And to get at it required that we remove the seat. That was done, and the fuel line was disconnected from the tank. When we removed the fuel tank we checked it out, drained the fuel and found out that at least half the liquid in the tank was water. Out of maybe 10 gallons, we had at least 5 gallons of water. No wonder the truck didn't want to start or run continuously. With 50 percent water, no engine runs well. And it all finally made sense. The tank, being in the cab of the truck, would get heated up to the cab's temperature, and air from the cab would get into the tank. All properly designed fuel tanks have a vent to allow air to get in. And this air would be humid from our breaths. When the truck was shut off and the inside of the cab cooled down, the humid air in the tank would also cool down and condense water out or into the tank. This sort of cycle had continued for a long time, probably even before we purchased the truck some three years before. And over time the water in the tank kept getting more and more. We could have alleviated the problem early on if we had put a quart of methanol or some sort of alcohol, sold as a moisture remover for cars and trucks, and kept up with the accumulating water in the tank. But we didn't do that, and when one has 5 gallons of water in a 12-gallon tank or so, alcohol wouldn't help.

So we drained the fuel tank and made sure it was empty, at least as good as we could get. Kurt was meticulous and got the fuel tank back to its original state, free of water. We buttoned the engine and fuel system up and fired the truck up. It ran like new, and that was satisfying. We needed to make sure to put in a can of alcohol at each fill up of gasoline, and we did.

At this time the children were all attending the Bonanza school during the weekdays, except for Sunshine who was 16 months old and was with Joy. While we worked on the truck we were treated to some of Joy's cooking at her "second home." It was appreciated, but I kept getting strange vibrations from both Joy and Ralph, and I too was wondering what was transpiring. I just kept going and didn't even want to consider the possibilities.

My ribs were healing up nicely, and the activity of rebuilding the truck was perfect for me. Kurt did all the hard work, and I kind of just paid the bills for the parts, made some suggestions, did some of the lighter work, and provided him company. I could visualize traveling to Montana on a bus. I checked the airlines, and it was no surprise that it would take longer to fly down to Butte from Dawson Creek than it would to take a bus. The airline route was circuitous at best, flying to Grande Prairie, Edmonton, a day overlay there, flying to Denver, Colorado, then to Helena, Montana, and a final jump in a whatever to Butte. And it was going to cost maybe three times a bus ticket. Montana Tech was going to pay for the fare, but it still was an easy decision.

And there was also some scuttlebutt in the area, and Joy was told by someone, whoever, that Butte was a mining town and a place one wouldn't want to raise children. I found this a curious attitude that somehow mining was like maybe leprosy or even worse. Were the people really any worse than the loggers, the big-time businessmen who seemed to have no aversion to taking all they could get, or even the religious folks who were always quick to judge others. I was going to go, and if I got an offer for the position, so be it. It was a long way from Baytree, and I had to figure out some way of doing that; would we all move down, would I stay there during the week and travel back and forth, or would we desert our dreams and return to the mundane at best?

So, about a week into May, I boarded a Greyhound bus at the Dawson Creek bus station. Probably it was a cool morning. The bus went to Grande Prairie with a stop or two in between, kind of like a local. The Greyhound buses were the only public transportation for many of the local communities, and often a small trailer was attached to the rear of the bus for small packages and parcel items. This bus took the day to arrive at Edmonton at probably near midnight. And my ribs were giving me messages that I should be careful, but how could I? I had a single seat, and the buses were only a bit better than an airliner. I do remember the Edmonton bus terminal as being rather busy with comings and goings east, west, north and south. My bus must have been in the terminal for an hour, and we left maybe around 1 a.m. heading south for Calgary and other points south. I think I did change buses in Calgary and south through Lethbridge, Shelby, Montana, and Grand Falls. My sleeping was mostly because of exhaustion, and even then it was a troubled sleep, with the pain from the ribs always there. We went through Helena, Montana, and over a rather significant mountain ridge. Dropping down, we arrived in Butte, maybe 24 hours after I had left Dawson Creek. I was met by one of the faculty members of Montana Tech, and they took me to a motel where I could get some rest and shower or bathe. The ride down had been very grueling for me, but now that I was not on the bus anymore I felt a bit better. I was still sore and would be for a long time, but I could maneuver.

The next day I awoke early, mostly because even now I was having trouble getting comfortable to sleep. But I was also getting pumped up to meet the Montana Tech faculty and see what was going on. I remember Butte, Montana, being not a very picturesque city. At one time, 30 years before, there were over 100,000 citizens in Butte and it was probably the largest city in Montana at that time. It was prosperous and mostly because of the copper mining that was going on there. Also, about 30 kilometers to the west of Butte was a copper smelting plant, personified by a large tall chimney to disperse various particulates into the air. It was an Anaconda Copper Company plant, and I am sure was a typical company, trying to get by without doing anything to help the environment unless they were required to by law or edict. As a consequence, the hills around Butte to the west toward the Anaconda plant, and even some to the east and north, were barren, poisoned by the belching of the chimney for the past 40 years or so.

I walked around the town, which now had probably 25,000 citizens and had lost about half of its land when it had 100,000 citizens. They lost this land because it was just eaten up by the large open-pit copper mine right there at the town. So while Butte was still getting by, it was losing ground, literally and I think figuratively. But there were some really interesting buildings, old stores and old homesteads that let you know that, yes, Butte was once the place to be. I think it had an opera house, a concert hall and a number of other establishments that indicated a prosperous and booming past. I remember walking to the railroad freight yard, the Milwaukee Road, which was the big railroad serving Butte, and I just rather wandered along like Henry David Thoreau would maybe have done. Henry would not have been in Butte, but I was just sauntering with no real destination, just taking in the scenes.

Montana Tech was a university of probably 10,000 students, mostly undergraduates, and it emphasized engineering. They were very active on environmental issues, even though the city and the outstanding industries, Anaconda and the mining companies, seemed to be against such efforts. And the faculty and staff at Montana Tech were friendly, accommodating and interesting. They were indeed serious about energy problems and considering real ways to solve them. Solar and wind were a big part of their mix, and I felt comfortable with the people and their ideas. I thought we would be a good fit, assuming they were being straight on with me. I have never been one who would tailor an interview for a position to suit the potential peers. And that probably was not always in my best interests at the time, but in the long run that is the way it should be. I filled them in on all my past experiences, and I often wondered what they actually thought of my excursion to Canada. Yes, even now as I write this, I cannot express the whole reason for the venture. Maybe it has to do with a concept or phrase I recently read

die Kunst macht die Furhbar Schrekliche Wahrheit Leidlich

The Arts make the awful truth endurable.

But my interview, from my perspective, was very positive, and I think I decided that if they were to make me an offer I would accept it.

I was treated as a guest at one of the faculty member's houses, just west of Butte, and they provided me with a good meal, a casual drink and some more good conversations. I think maybe two or three other faculty were there as well. And I remember thinking that this was a good format. Too often a potential new faculty member is taken to some local restaurant and provided a nice meal and conversation, but that is always a sterile atmosphere finally. Homes are where things really happen, and this is too much forgotten.

Probably the only downside to Butte to me was the barren hills to the west and north that had been polluted by the Anaconda plant, and that really bothered me. What would this place look like in another 20 or 30 years? I would hope that the copper would run out, Anaconda would move away and the land would recover in some fashion. And the evidence of acid rain and lead, mercury and various other trace elements falling on the rich earth and killing the trees and ground organisms and microorganisms was enough to make me take a real hard look. But I think finally I have always been optimistic about the earth, not the world of people but the earth. There must be some human beings out there, maybe on some forlorn Indian Reservation or in the jungles of Africa or South America, and some meek beings who will inherit the earth, not the world....

I think I spent two days in Butte, the first day and evening with faculty and staff and the last day with the university president, the dean of the college, and probably a personnel officer, explaining the benefits and salary configurations. It was all rather an expected format, and it all reminded me of the University of Dayton, Wright-Patterson Air Force Base, Monarch Machine Tool Company, and Perfect Circle Company. Even my high school and university summer work for a couple of summers at Overmyer Mould Company. It all seemed so silly, really. And the people, poor people, were so absorbed in following the right format and making sure they conveyed something to me. My almost two years in Canada had somehow given me a reality check, although I was being negligent in not fully integrating my family into my thoughts. How could I when I didn't even know them myself? We all seem to walk around in a dream, an illusion, rather stoic, yet with full *angst* of what might happen. We know what will finally happen, so let's get on with our lives.

The next morning I boarded a Greyhound bus for Edmonton. It retraced my journey down, and I was again reminded of Morning Star's impression on me. My ribs didn't like the jostling around of the bus. And sitting in a bus seat is just not comfortable, only marginally better than the airliners at that time. Since that time, the airline seating is almost beyond or maybe below description. How can one sit in a seat not as big as child's footstool for an hour or more? My brain just rebels at that and also the realization that most everyone I know accepts it as part of travel. Almost like the poor Jews walking into the gas chambers and not knowing what is awaiting them. And I know I must be considered a total reactionary impractical whatever.

Again, I rode that full day and arrived in Edmonton late at night, the bus terminal this time being a bit dismal but still busy. We then proceeded toward Grande Prairie, and I arrived home the following morning at Dawson Creek. I am sure I called Joy at home to tell here when I would be home, or maybe I talked with Kurt or Tim or Lori. And they all knew beforehand my proposed itinerary, and there was someone at Dawson Creek to pick me up. I was glad to be home and with my family. Then two pains came to the forefront, as if the broken ribs were not enough.

Dental treatments in Canada were individual responsibilities. And that meant Joy and I weren't good about going to the dentist for cleaning or checkups. I had recurring toothaches. Actually about this time they started to get continuous, and I finally just went to Dawson Creek and had a couple of problem teeth filled. That fixed my tooth problems, and I thought at the time how nice it would have been to have my rib cage just kind of patched up quick and neat. But that kept going.

And the other pain was that when I arrived home Ralph Tompkin's road grader was parked in our yard. He had offered to get our lane graded better. Why that? I didn't like it or want it at all. The grader was to remain parked there for a few days, and on one of those days Joy sheepishly told me that Ralph wanted her to divorce me and live with him. That was his pattern when he went to Utah to find a wife and now again. I felt really low, like after 19 years of marriage I had to start out all over again. I was stunned in a total way and wandered around for a long time, even after Ralph came and took his road grader home. While it was there by us, I was tempted, even rather went through the thought process of putting sugar in his gas or diesel tank, intending to ruin the engine. I came very close to doing it but somehow didn't do it. I would be stooping to his level. But I sure knew then how King David's (from the Bible) Captain Uriah the Hittite must have felt and thought.

So I had plenty to keep me busy. I did try to help Joy in her struggles, but what could I do, or maybe what should I have done much sooner? I am sure I had caused the compromise as much as she.

About a week later, I received a telephone call from Montana tech. They wanted me to join them on their faculty. That was a very good bit of news that carried me through all the worries and pains that had come around. I think I may have told them I would let them know in a day or so, but I knew what my answer would be. And even with the rumors in the neighborhood of the problems of raising a family in Butte, Montana, it all just seemed to be inevitable. My dreams were become faded and foggy, and I felt like maybe we could hold on to the farm here and I could spend the school year in Montana and return for farming in the summer. That was mostly wishful thinking, but one does need to be wishful from time to time; the Arts don't always help.

CHAPTER TWENTY-SIX
Home Improvements

My commitment to teach mechanical engineering at Montana Tech in Butte, Montana, changed all our plans. Montana was a beautiful state, even though it was a state and even though Butte was in a polluted hole. We would be close to Yellowstone National Park, and it would be a day's drive, hard day's drive to Baytree. We would see. And the plan to build an addition to the house was now something that seemed to be a good idea. It might make the homestead more attractive in case we were to sell out. And I had lots of building material, wood, that would give us the major parts of that addition.

The lumber we cut with our allotment and milled with Jess Peachey was dry piled and completed on April 11. The weather was getting a bit warmer now, and spring was hinting to us. Even with the plans for me to teach at Montana Tech, we went ahead and did the farming we had planned on doing. We had a 20-acre or so hay field that was the result of the previous spring's oats and hay seeding. I did seed about five acres to wheat that we could also use as food for ourselves. We had found that cooked wheat was tasty and nutritious as well. And wheat could always be used as animal feed for pigs or cattle. Horses didn't really care for it. The wheat was seeded on April 30, and the next day was warm and sunny and got up to 85 degrees F (30 degrees C). I think the wheat must have liked that because it sprouted nicely in a week.

On May 6 Flossie our cow had a healthy bull calf, and both seemed to be fine. The weather suddenly became like summer. It did rain a few drops one day, but it was getting dusty and windy. The highway crews were surveying Highway 49, getting ready to put down a pavement. It was ironic, because one of those early May days, while it was almost summer-like, I drifted over to Highway 49 and just sat down in the new sprouted grass along the edge of our woods, near the corner, Pilsworth Corner. The highway was now rutted dirt and pits, not smooth like it was during the long winter months. The few rains and the melting snow had put those ruts in it along with lots of heavy trucks and even the pickup trucks and cars traveling at excessive speeds. Somehow I felt good about doing this. We were working nonstop, nearly since coming to Canada, and now I felt an odd sense of relief mixed with regrets and disappointments. It was sunny and dry, and suddenly an automobile traveling east from Dawson Creek, probably, stopped directly across the road from me, the driver rolled down the window and yelled at me, "You have a hell of a nerve calling this a highway." It was an interesting comment and one that typified the age, I think. Even in Canada people were always expecting paved roads so they could travel at high speeds, and rutted roads were just an insult to them. If the driver had come around another month later, he would maybe have traveled on the final paved highway, but at that point it was still the top-layered dirt over the initial grading.

I didn't say anything as the driver rolled his window back up and headed east toward Gordondale and Spirit River in a whirl of dust. Good for him and them.

And sometime in late spring, with the children still in school, Joy and I and Sunshine went to Moonshine Lake during the day, just for a break. It was a sunny day, one that made the Peace River Country at that time so special. The air was clear and smelt fresh and new; the trees were relishing the warm weather and the slight breeze brushing their leaves and limbs. It was done rather unexpectedly. Joy and I had lots of mental and verbal bruises from each other and our behaviors, and it was one of the few times that I remember we did something together, just she and I. We did have Sunshine along, and that was good, but I think we wanted to somehow shore up the relationship between us. It was nearly shattered, and the fire was down to a low bed of coals, and this was an attempt to bring something back. I remember Joy had a nice quiet comfortable blooming pants and jacket combination on. She was beautiful and at that time comfortable to be with. The lake was quiet and therapeutic. No one was there at the park next to the lake except her and me and Sunshine. There were some swings and a slide and maybe a merry-go-round. Sunshine went down the slide a few times and sat in the swing. It helped us and was something we should have done at every impulse, but we didn't. It was only this one time in Canada, and I almost felt like it was back in 1960 or 1961.

From here on we would work even harder than we had for the first 21 months or so. We had to do a full summer's worth of farming and haying, build an addition to the log house, and keep the horses worked and gentle. I had gotten back to handling Morning Star and Bess again. Harnessing the horses was a real chore, but I managed, with help from Doreen for a while. We didn't have any more hay to haul, but we did haul some logs down from above Pilsworth, mostly for firewood and maybe an occasional pole for the woodshed or whatever. And then too we were looking at a possible move. Actually we were planning on moving to Butte, and Joy had quit her job. I didn't get all the details, but that made me feel much better. We were never right to the last penny on our funds, and we were looking at a steady income in four months. Back to somewhere, eh?

In the middle of May we brought a big planer over from Jess Peachey's. I am not sure now if it was his, whether he had rented it or what. But it was a big machine, and our John Deere D was the candidate for pulling it. Someone suggested it was not big enough to pull the planer or that it would not pull evenly enough since it only had two cylinders. Again, the old John Deere had to prove himself and he did, very well. With our big thrashing machine flat belt hooked to the planer and the D out a good distance, it just drove that planer like a little red wagon. Well, not a red wagon, but it made the planer hum.

A planer of this sort has two rotating arbors, each having a cutter or sharp blade that smooths wood board as the boards are fed through the planer, between the two arbors. There was also a series of feed rollers that would actually push the boards through the planer so that all one needed to do was place a board on the table at the head of the planer, the feed rollers would literally grab the board and feed it between the rollers. The noise from this was hard to bear. I think we all had ear plugs or ear coverings to protect ourselves from the noise.

As in milling logs, many hands made light work, and I think probably Kurt, Tim, John Trinier and maybe Jess Peachey planed our boards over two days. We had a large amount of boards, and it wasn't doable in one day. And it was good work. We would trade with Jess to help him get some boards planed after this session. And as just another hint of something or someone watching our whatever, once I was standing at the side of the planer next to the big flat belt. The belt was moving at maybe the equivalent of 15 kilometers per hour and all at once a board was dropped somewhat off the table, it landed on the belt and came whipping down toward me and actually hit me in the upper leg. It gave me a small bruise, but it all could have been much, much worse. I always seemed to be an accident looking to happen, somehow. And of course, my ribs were still tender. But we now had a pile of planed lumber that would be good for the addition.

Joni Byler and his wife moved up to Baytree in the spring of 1980 and bought a half section of land up on the hills to the east and north of us. These were not the Saddle Hills but a rather low ridge of soft hills, and they had a nice setting looking to the west toward Dawson Creek. They had done a lot of moving and traveling during their lives, and this was just another one of them. Joni was a Mennonite bishop, so the Baytree church now had two, Willard and Joni. But Joni was such a settled, mild, kind person, and he never even implied a sharing of the "power" with Willard, at least not during the time we were there. The Bylers as things went would only stay in Baytree maybe three years, and they too moved away, to northern Michigan round Baraga. But that spring and summer Joni volunteered to help me do the addition. He was skilled in many vocations, and one of them was concrete work, something I had no good experience with. I could cobble a footer and probably even do a rough sidewalk, but I claimed no skills at smoothing a surface of concrete. This summer would change that a bit.

Jess came over one day and dug a basement for us. Somehow we seemed to think we could use a basement. It was self deception again because no one had a dry basement, so most people just didn't have a basement, period. But we felt, Joy and I, that a basement would be added room for the house, so we had it dug out, about 2.5 meters deep, the width of the house, and 20 feet out. I sketched some possible room arrangements and thought a two-story addition on top of the basement would be good. Talking with some of the neighbors, Ian and maybe George, and George FitzPatrick, it seemed that using treated plywood for the walls would be the way to go, least expensive, and the treated plywood would hold up for a very long time. We had a cement mixer, probably Joni's, and we got in a load of gravel and a few bags of cement. One day we started, Kurt, Joni, John Trinier who was Joni's son-in-law now, and maybe even Jesse Peachey, who was always willing to help out and who we owe so much, mixing concrete, loading it into a wheelbarrow, taking it to the basement excavation, and pouring it in. We poured the floor that day, and with lots of hands it went smoothly. Joni was the expert concrete man and knew how to level and smooth the floor. I am not sure if it was my idea or Joni's, or Kurt's, or whose, but we put a small pit at the corner near the log house side, about 30 meters deep for a sump pump to keep the basement dry. Sump pumps are designed for just this sort of application, removing water from floors and basements and "sumps." We put the pump in the small pit so its intake would be below the basement floor level, and that could keep water out, maybe. But for the pump to operate it would need electricity, and we had no electricity. We did rig up a battery for a while.

During this time we had a few trips to Joni and Liddie Byler's place north and east of us. It was rather up on a small hill, and looking west was a spectacular view of British Columbia and some faint glimpses of the Rocky Mountains. His property was wooded to the east, and they had hurriedly gotten a mobile home for a temporary dwelling. They were planning on farming, and with their expertise, I had all the expectations that they would succeed. They just seemed to know how to go about setting up a new farm, unlike myself. One day, after visiting with them for an hour or so, we headed home in the pickup truck in the late afternoon. In the summer, late afternoon was still filled with another four or five hours of daylight. So it was sunny and bright, and as we headed down the lane toward the road we saw a brown bear, a sow with her two young ones meandering in the fields perhaps 500 meters to the west. They were grizzlies, and the sow was big. It was just another sight that one does not get to experience in very many places in the world anymore.

Now with the basement floor poured, we could go ahead and erect four walls with 4-foot-by-8-foot 3/4 inch-thick treated plywood. This all went smoothly with all the hands for a second day, and we progressed quickly with the basement of the addition. We put a plate, 2-by-8 boards, around the tops of the walls for a support for the floor joists of the main floor, 2-by-8 boards again. I remember that we covered the floor or deck with inch by 6-inch boards crosswise at 45 degrees. That would stabilize the basement walls and keep everything square on top. Ian told me, after noticing that the wood was all spruce, that spruce shrunk a quarter inch per year. Well, at that rate, a 6-inch-wide board would be gone in 24 years, shrunk away. But the deck looked nice when we had it all installed, and even just after installation, one could sense that the boards would shrink; there were some small cracks between the boards.

I felt good about accomplishing this much. We now had a substantial deck as wide as the log house and extending out 20 feet and raised level with the log house first flooring. That put the deck about half a meter above the ground or even a bit more. As things finally all continued, I didn't find time to frame up walls and a roof; it was just the deck and a basement under it. I did put in a staircase from the deck to the basement floor, a two-flight one, with a left turn halfway down. That worked out nice.

And I kept at getting work for the horses. We did keep getting some firewood, dead logs, from the hills south of us, and that was relaxing and satisfying. I wondered what would become of my dear horses if I sold them, because I couldn't imagine taking them to Butte, Montana. There was an outside chance for that, but who knew. And I had brought along my hay loader, hoping to use it to make loose hay sometime. That seemed to be a fading dream now, but I did go ahead and take the hay loader apart where I thought it would make painting it more convenient. It was mechanically in perfect shape but did have lots of faded paint, so I sharpened the paint up, and I did it all down near the Twelve Mile Road. There were a few people who stopped and asked about it. It was an imposing piece of equipment that the natives would probably not have seen for 40 years. I think when the farmers in the Peace River Country made loose hay they used a hay sweep, kind of a large platform that horses would be hitched to at the rear, and the team of three or four horses would literally push the sweep in the field of hay, stacking it up and then backing up and getting a run at another patch of hay. The hay loader was most suited for loose hay loaded on a wagon and then taken to a large cavernous barn to be stacked inside. Here the hay often was best left outside.

And all the while I wondered if there were any people needing hay. I had a big stack of hay that I just couldn't move. And with some luck I was going to get some more hay this summer. Oh well, it was like money in the bank, eh?? And we kept right on making hay as if we were putting money in a savings account. The Peace River Country was good hay country, and we took advantage of that. We used the John Deere D for pulling the baler and a flatbed wagon to stack hay as we now knew that just stooking it in the field would not do it. And one day Kurt and I were getting ready to bale some hay in our field. The baler was attached to the tractor and the wagon behind the baler. I remember that Kurt and I tried to make sure everything was in order, and actually he was more precise than I. I was on the wagon, and he was checking out the power takeoff to the tractor. As events unfolded, he had the baler in gear so the power takeoff shaft was rotating. Suddenly I saw Kurt standing by me at the wagon, and he looked ashen and his eyes were dilated. I looked at him and then I saw it; his right pants leg was gone, ripped off his body by the rotating power takeoff. We had been using our clothes hard during the past two years, and his pants were worn thin. When the power takeoff grabbed his pants leg, it just ripped away from the remainder of his pants. If he had had new firm, strong pants we probably would have lost Kurt as he would have been whirled around the power takeoff shaft. Just such accidents are recorded very year, and we surely were so fortunate, and as the events folded out, he was still together. Kurt changed into another pair of pants, and we proceeded to bale hay, being more wary of the power takeoff shaft than we had been.

And during this time we got a litter of piglets from our sow and that, after we had purchased a few small pigs. So I think we had a sow, maybe seven piglets, and then five growing pigs. It was getting to be a real menagerie. Prince, Bess and Morning Star were doing just fine too.

We also had not lost our bearings completely. We needed eventually a refrigerator or place to store and keep food cool. In the long term I thought a root cellar would do us well, but that right then was a low, low priority. So we bought a gas refrigerator, an absorption Electrolux-Servel machine, from Bill and Nelly Gamble up Twelve Mile Road. I think we paid $60, and we brought it home and put it in the log shop. We would need a propane tank or kerosene fuel, I cannot now remember. But we did not use it very much, and it mostly just sat waiting for us to want it and recognize its usefulness.

And we became befriended by Gordon Jones, from near Gordondale. He was a teacher in Dawson Creek and traveled back and forth past our place. Gordon was a very outgoing, caring person, just what one would expect from a committed teacher. And he got to coming around after school and when he went to Dawson Creek for errands. He sometimes just stopped and talked, but sometimes he would have a tub of ice cream; my daughter remembers root beer float ice cream. Our children must have known that, yes, we were now in Heaven.

And then, sometime in late July, I received a call from the University of Wisconsin-Platteville. They now decided that they needed another mechanical engineering faculty member and wondered if I was still interested. I told them I had committed to Montana Tech, even signed a contract, but that I would come down. Wisconsin had a reputation for high quality education and a good school system, but yet it was odd that they told me they would pay my travel expenses, but if they made me an offer and I refused, I was obligated to repay those expenses. It seemed to be a really cheap shot, not professional at all. But I did see where they were coming from; they didn't want to finance travel for someone going out on a lark, so to speak. Oh, well, that put Wisconsin down a little in my estimation, but we would see. I would do some more traveling before long. I think it was the third week in July that I would travel to Wisconsin.

CHAPTER TWENTY-SEVEN
Decisions

The second summer in the Peace River country was speeding by. We had some rather hot weather, there were a few nice rains, and the hay field was looking good. We could probably cut the hay and bale it in the last week of July or the first week of August, depending on the weather, whether it would be dry or whether it would be rainy. But I also had to travel to Wisconsin to interview at the University of Wisconsin-Platteville. Even though I had accepted a position at Montana Tech, I decided that Wisconsin might be a better place, maybe. It was closer to Indiana where our parents were and also more like the climate we had grown up with.

Then on July 4 Joy's father died quickly. He was having some health problems, phlebitis in the legs and some blood clotting in other parts of his body, and so he was taking some blood thinner pills and some blood pressure pills and other stuff too. Not a good thing to take pills for anything but more than one is asking for big troubles. He was a farmer all his life and was not good just sitting around. So one day he climbed up on one of his tractors to do some work, or whatever, and he put the tractor in reverse gear and had a massive heart attack and slumped over the steering wheel, dead. The tractor went in reverse until it backed up against a building, and his son, Joy's brother, found him on the still-running tractor. Joy immediately made arrangements to fly down to Indiana to attend her father's funeral. She would take Sunshine along because she was 19 months old and still nursing. She flew out of Fort Saint John, British Columbia, that evening on a Canadian Pacific airliner. I took her to the airport and even took her into the airplane's cabin to help her get settled. Joy, up until then, had not flown on a commercial airline, so she was a bit uncomfortable. I remember that there was some mellow background music, not hard rock or other noise. I felt lots of sympathy and empathy for her as I had lost my mother in 1974, and that was hard to handle. She and Sunshine were off, to return about four days later. We would hold things down at the farm.

I was reluctant to mow hay in mid July because I was committed to going to Wisconsin the third week. So during my spare time I was inventorying our property, equipment, tools, furniture and everything. I think I had a mental block about Bess, Morning Star, Prince and even Flossie. And I did make arrangements to fly to Platteville. Since it was not the immediate notice type like Joy had with her father's funeral, I flew out of Dawson Creek, to Grande Prairie, to Edmonton, Winnipeg, Chicago and finally Dubuque, Iowa. The flight out of Dawson Creek was late morning, and with a layover in Grande Prairie, I arrived late in the evening in Edmonton. I had to stay in the airport terminal until morning, and I remember that night with cloudy glasses. It was a surreal evening. The sociology, technology and general people behaviors were moving and shifting quite a bit during this time. In 1980 the world finally received a rude shock that the oil was eventually going to be gone. The Arab world became rather entrepreneurial, or capitalistic, or selfish--all the same. They decided to unite into an economic block and decide what price oil should be and practice an oil embargo to enforce that concept. They actually were doing this for a year, and in the US the embargo was a rude awakening. I am sure many of the US citizens wanted the government to go to Saudi Arabia or wherever and just take it over; after all, the US had more weapons and will to use them than any other country in the world, even Russia. Somehow, saner minds prevailed. The US President at that time was James Carter who was not so belligerent as the next President would be. But we heard stories of people waiting in line at a gas station for an hour or more to fill their fuel tank in their car or pickup. Yes, even the US citizens had pickup trucks. So gas was now known to be finite and also political. But also the people were becoming more selfish, if that were possible, and less concerned for their friends and neighbors, just what Nietzsche and Kierkegaard predicted and would have expected. And the music was becoming even stranger and more ugly with time. The airport terminal, the lounge area at Edmonton, was new, and there were lots of benches but few comfortable chairs and surely no sofas. I was obligated the sleep on a bench in that terminal, with dim lights (they at least turned down the lights after midnight), and weird background noises. Some--the aliens walking around, or maybe they were people--might call it music, but it was not. I was still feeling the kiss from Morning Star and the kiss from Joy too. It just all seemed so foreign or rather alien. I did not sleep well, very troubled, and I also did not want to walk around too much because there were just enough people wandering around to make me be suspicious. Probably unwarranted.

The next morning I boarded a flight to Winnipeg which would take maybe two hours. It was a typical airline flight, boring, noisy and uncomfortable. They did provide a snack and a drink. At Winnipeg I went through US customs, wandering up and down various flights of stairs before I was back in the waiting area of outgoing flights. I probably waited two or three hours at Winnipeg before I then flew to O'Hare airport in Chicago. That too was a jumble of chaos and ugliness. Not at all human. But I transferred to a smaller two-engine jet that went to Dubuque, Iowa. I remember flying over the Mississippi River and feeling a distinct bump, like we had gone over a ridge. We had gone over a thermal air current from the river below. I had experienced this same phenomena back in 1976 when I flew to Utah to visit with Tracy Hall about my high-pressure physics work. In a way it seemed almost like returning to Dayton, but not. The aircraft landed at the Dubuque municipal airport, about eight kilometers south of Dubuque, where I was met by Ross Fielder, who was the chair of mechanical engineering and with whom I had talked on the telephone at home. He was a very friendly and outgoing person and seemed to be glad to see me. It was a warm sunny day. I could tell that I was not in Canada now. In Canada, even on a sunny and warm day there was still the hint that winter was coming. At 90 degrees F you knew it would not last. It was drier and somehow cooler than this temperature in southwest Wisconsin, northeast Iowa. As we headed east and a bit north I could see ahead a large mound, almost a mountain, that had a large white "M" on it. That seemed strange, and I must have thought some about it before asking Ross what the M stood for, was it an upside down W for Wisconsin, or what? He filled me in that it was for Mining, or Mining Engineering which was the first engineering program at Platteville dating to the early 1900s. Interesting, and he also told me that it was the largest M in the world, yeh, right. Oh well, maybe it was, but I never was impressed with that glob on a beautiful hill.

Southwest Wisconsin was rather hilly, or rolling. Farther north one could see that it became distinctly knobby and very agricultural, rural. There were lots of corn fields, hay fields and cows grazing in pastures. It was less populated than Ohio or even Indiana, and I liked that. It was not the Peace River Country but not over-populated either. Here there were no lakes but many rivers and creeks. Platteville was about an hour from the airport, and we arrived in the late afternoon. I would stay two nights in a sprawling motel called the Governor Dodge Motor Inn, kind of a colonial architecture yet typical motel.

The following morning I was treated to a good breakfast with Ross Fiedler again, and he gave me a tour of the campus, not a big campus but not small either. It covered an area more than the University of Dayton where I had taught for 12 years. And in comparison, the buildings on campus were just generic boxes, seeming to be solid and well-maintained but still boxes with no architectural values. There was one building that looked interesting, Old Main, but he told me it was scheduled for demolition within two years. And he showed me the original mining school building, Rountree Hall, which was an interesting limestone building having a tower and lots of fascinating nooks and crannies. It was vacant at the time, having just been moved out of by the engineering school a year or two before, so they could move into Ottensman Hall. The University of Dayton had most buildings that were over 100 years old. My office was in Saint Joseph Hall which as a beautiful old five-story building with lots of dormers, a busy roof line and towers at each end with statues of Joseph and Mary in alcoves of the walls. Platteville was surely not Dayton, but then again, what was I to expect? Ross Fiedler was the lone faculty member I met, although I did meet the secretary, Joyce Clifton, and the mechanical engineering technician, John Abing, both friendly and seemingly straight, small-town people. All three had only been on campus for three years or less; John actually had started that same year and Joyce two years before. Ross came three years before. And the mechanical engineering program was just two years old at the time. The university also had three other engineering programs, mining, civil and industrial, and they were hoping to establish an electrical program within five years. So it was a growing program.

Ross explained the goals and educational philosophy of the programs as being practical engineering programs. It sounded exactly like my educational experiences at UD in mechanical engineering technology, and I wanted to expound a bit on that but decided not to. I was qualified and felt I could very comfortably contribute to the university here. The mechanical engineering program did have six other faculty, one a new faculty who came on just the same semester. So I would be the seventh faculty member. Ross explained that he had just taken over the chair's position as the previous chair, James Scroggins, had rather unexpectedly left in midsummer; thus they had another opening, and I was the first candidate for that position. Also, the dean of the college of engineering was out of town, so Ross was the point person to see if I was acceptable for the opening. The dean would have the final say, although he would be taking Ross's word for it.

Ross explained again the policy that they would pay for my trip unless they made me an offer and I refused it. Then I had to pay for the trip. Oh well, insanity is not rare in institutions or groups but rather common. I was taken back to Dubuque the following morning to board a flight to O'Hare, Winnipeg, Edmonton, Grande Prairie and Dawson Creek, retracing my journey down. I was glad to get back home and enjoy the remainder of the summer.

We cut some our hay field with the 6-foot horse mower and raked it into windrows. And we baled it all with the John Deere D. I could visualize doing this all for the next 40 years if need be. It was getting comfortable and only after a little over a year's experience. But I was not able to generate a steady income, even a small income. My savings had dwindled down to a point where I just had to get some financial income to support the family. Somehow Joy and I went over the bump, and she was not doing the housekeeping any more. And that was a trade-off because she did make enough money to cover many of the expenses. And the family was growing and would need new clothes, more food, and more room to live in. I kept going around that bush and always came back to just going back to teaching mechanical engineering. So my position in Montana Tech was a material comfort to us, if a desertion of my dreams.

A week or so later I received a call from Ross Fiedler, and he made me an offer to teach at Platteville that fall. I remember the salary being slightly more than Montana Tech had offered me, and I indicated that I would get back to him soon, knowing that I had to pay for my travel if I declined. And the only question, ethically and professionally, was what Montana Tech would think. So I contacted them and explained the situation to them. They were very accommodating and seemed to understand. They were willing to release me from my obligation. So I accepted the position at Platteville.

After this all sank into my mind, I realized that Platteville was too far away for commuting and even too far to expect to spend summers at the Peace and teach in the fall and winter. It was clear and brutal that I had to clear everything out from our dear home and barn and land. Somehow the mind can work in strange ways, like making a hand numb if a finger is cut off or some other absolutely painful injury. Trying to handle the thought and the reality that I was going to lose my dream seemed to have gotten unconsciously pushed way back in my mind, and only now, some 30 years later, is the pain of it coming back. At the time I went through all the necessary motions to get ready for moving to Wisconsin, not reflecting anymore on the dream. And this included planning for an auction to sell off what we couldn't take with us. There was no boxcar this time to load up and take south, no promise of a new land to explore, nor a challenge of planning a new life. We were going back to a straight planned regular existence.

Looking ahead, when I came to Platteville, one of the faculty, who was in charge of finding speakers for a monthly forum and organizing such presentations, found out about my endeavors and asked me if I would give a talk about our living in Canada. It sounded interesting enough and whatever, but I was still in that mentally numb state and maybe something like a denial state, and I just told him I would think about it. That was over 30 years ago, and it is now strange that I think about that time and know it would have been just way too much for me to handle at that time. It is hard enough now....

CHAPTER TWENTY-EIGHT
The Auction

During the second week in July we arranged to have an auction during the first week of August. Posters were placed all over the region, I think even Fort Saint John and over to Spirit River and Rycroft. I am not sure if the distribution went as far as Peace River or to the southeast, Grande Prairie, but this was probably the sort of advertising I should have done to try to sell my hay. So now I really had to get ready for the sale and complete the inventory of our sale items. It was not fun at all. Two years previously we had an auction sale in Ohio, and even then it was kind of hard to part with some of the items we had, but at that time we were looking to go to a better life and a new country. This time it was a regression, and everything was just going to be gone.

We kept busy making the hay and stacking it. I wondered how the hay would sell at the auction. I just hoped we would break even. And the weather in early August was warm, rather long days, and sunny most of the time. We were still going to church now and then, but I think it may have been the first Sunday in August when we attended the Baytree Mennonite Church for the last time. They had moved into the new building by Jim Sadlier's, and it was all bigger and newer. Before the services I remember Joy and I were standing outside, and Phil Sadlier was there too. Everyone else was inside, but we got to talking about everything. Phil was kind of earthy, at least edging that way. He must have known our struggles, at least with the church, and he tried to make the conversation one of encouragement. But of course he fell into the baseline of the churches: "God loves you and forgives all of your sins; just say that you believe and you will go to Heaven. And, yes, we are all hypocrites, but there is always room for one more in the church."

My thoughts were that that was what we had hoped for in moving to the Peace River Country; going to Heaven. It was not to be, yet. And I had gone through so much during the past two years and was still enduring it that I didn't feel up to attending churches again. Joy may have had other thoughts, and I think Kurt and Lori were still church members at Baytree. I would always respect their ways and thoughts and would never tell them that going to a church was not the way to Heaven or even a satisfying life, but those were my thoughts.

And every other day or so one of the neighbors would come visit with us. Those times would vanish in the Peace River Country as well, when neighbors visited one another. During our two years there it was becoming evident that the outside, southern world of people was changing the sociology, and neighbors were becoming just unknowns. We were probably leaving at a good time. Werner came over one day and told us he would be at the sale. George and Arlene Blakus too came over, as well as George FitzPatrick. One afternoon Ian came bouncing up the lane in his truck, and we sat for an hour or more on the deck. It was sunny again and warm enough. Ian was associated with Ralph Tompkins through the oil business, although Ian himself was only a sideliner, a subcontractor, a flunky. He knew Ralph, and in a vein where we were talking about Joy's housekeeping job there, Ian made a comment that I recall, just now. "Yeh, Ralph is real Ralph, isn't he. You can't trust him." Somehow I think Ian knew as much or more than I did about things. It was a large area, the Peace River Country and even Baytree and Bonanza and Rolla and Dawson Creek, but there were not that many people, and gossip and juicy news were the first to get around, and it was like the viscous, tenacious dirt roads; it stuck. He must have had a lot of information, most probably not true, but then, what constitutes truth, eh?

The presentations at the auction would be conducted at the log house and barn for the animals, furniture, tools and small items. We also had taken the horse buggy, and it was at the house, along with the box-bed wagon I had built up from a wooden four-wheel chassis in Ohio. There was now a substantial pile of firewood that would go, Flossie and her calf, and of course Bess, Morning Star and Prince. Those three would be the most difficult, but at the time, as I said, I was mentally numb. The tractors, green wagon, hay loader, mower, rake, baler, pile of lumber, hay and grain would be sold down by the Twelve Mile Road, where the new granary was situated. I can't remember whether we sold the granary, because it was portable, but it was not on our listing of the sales at auction. I may have sold it privately, so to speak, later. I still had the thought that we could return to our home after maybe four or five years, but that was more than a dream or illusion. It would not happen.

And as the sale day approached it was becoming obvious that it was a significant event for the surrounding region. We did have a number of people come days before the sale to look everything over. This was not the normal auction sale for the Peace River Country. It was somewhat like an auction sale of 30 years before, but there were some odd items, the two John Deere tractors--the A and the B, the hay loader, and the buggy that had never been in the country before. And I could see that it was turning into an entertainment event as well.

We had good weather for the sale, sunny and crisp but not cold. There had been a dry spell before the sale, so parking was accommodated in our field by the granary. And we had people come from all over. I did not count the number of people, but they were all over the place, in the house, in the yard, in the barn, down by the granary and up and down the lane. And about 10:30 in the morning the auctioneer started off. Small items went first, and as I read the listing of the sale after all these years, we had a good sale. Auctions, to be successful, need to have at least two people bidding on the items. We must have had that. And not to get sucked up in greed and wanting more than items were worth, the sale was just a continuous case of getting more for an item than we would have gotten in Ohio. Our cook stove sold for $250 and the pot belly stove for $325. In Ohio these would have been sold for $10 each, maximum, maybe $10 together. Joy's treadle sewing machine went for $180, and in Ohio or anywhere in the US it would have been lumped with "something of value." We had so many items that in Ohio would have all been scraped together in a box and sold for a dollar bill. Here we were getting at least $5 for just a plate. I built a bench that served as a rough sofa, and it had a storage space under it that was identified as a Chesterfield and sold for $50. In Ohio it would have barely been firewood. It all became overwhelming to me. As I keep saying, I was more in a trance than with it all, although no one would probably have guessed it. I am a good faker too!!

One of my friends, I think maybe Peter Kut, came up to me during the sale, maybe at a lull, and told me "Kurt, this was a young man's dream that brought you here, wasn't it." Yes, that was certainly part of it all.

And I drifted through the day not really understanding any of it. I do remember thinking "Where were all these people when I was trying to sell hay, or eggs, or milk?" If they had come for that, we could have stayed and this sale would not have occurred. But the shadow of all those years back makes it less opaque now. We were leaving at almost the very best time, even though we would not know it for 30 years.

The washing machine we excavated from the house at Baytree sold for $30, which was one of the items that sold for a bit less than we purchased it for. My box-bed wagon brought us $240, and the hay hooks I brought thinking that eventually I would have a barn with hay mow for loose hay sold for $22.50. Again, this was scrap metal in Ohio.

The two-seat buggy brought us $1,650. Flossie went for $650 and her son for $450. We probably would not have gotten that at the auction barn in Dawson Creek. And I was there to witness all this, in my resigned state. Bess would be the hard one, and somehow I remember standing way back just hearing it all but not. She was taken by Evert

Parlee, who knew Bess well, for $3,700 and Morning Star for $5,700. In my mind he got a bargain, but others wanted the two as well. Ian Marchuk got the stud colt, Prince, for $2,000. The sow went for $120 and the piglets $200. I suppose they made good "hams and chops" up there. The Canadians are cannibals too, and there are very few people let alone Human Beings, who would save Babe.

The harness and collars and lines and assorted stuff brought a good amount as well. Financially, the sale was a success without going any further. But then we had the equipment and wood and hay and grain to sell. The auctioneer took a break of 20 minutes or so, probably, and some lunch. Joy and the girls had planned it and had a lunch stand selling soda, coffee, tea, sandwiches and chips, lots of typical auction lunch stand items. They did well on that as well, and I think they actually sold everything and could have sold twice as much. We knew there would be people coming, but we did not anticipate this many.

Then the auctioneer and his troop walked to the end of the lane, on Twelve Mile Road and down to the granary and the remaining items. The lumber went first, and the pile, mostly intended for the new addition, went for well over $4,000. That proved to me that if things were sane, we did have a viable path to a sustainable life in Canada. But it was all evaporating, or maybe boiling away. Our sweet 5-foot hay mower went for $110 and the stubby drill $325. Most of the sale items and the buyers of them involved entertainment and hobby stuff. But then agriculture was even here turning into agripornography, and the equipment and setting we had was an antique, maybe even an art, eh? The hay loader, which I had carefully and I think lovingly freshened up, like a pretty lady getting her face and body into a perfect presentation, brought $650. I had paid $25 for it, and now my feelings were, "I hope that person takes care of it," again a self deception, an illusion.

The horse trailer which brought Bess and Morning Star to the Peace River went for $2,100 which was fair, but then measuring things by dollars is the root of the root of all evil. The faithful thrashing machine got us $150, not bad... And as we circled around to the tractors, everyone was there to see what they would bring. The big John Deere D, which was more valuable and workable than two or three of the big four-wheel Versatiles or Steigers, or John Deeres, or International Harvesters or whatever, brought $550. The baler went for $575, and the John Deere A was $1,000 of filth. As the crowd approached the John Deere B, a man was sitting on the seat and casually had placed his foot on the flywheel. I know he didn't know what the flywheel was for, why should he, it was not new??? And he just moved his foot forward a bit, kind of reacting and wanting to not miss anything, and the old girl took off. The engine putt putted like I had known her to do for the past 10 years or so. The man had turned over the flywheel maybe 10 degrees, and that was enough to ignite the fuel in the combustion chambers. I could see the reaction among many in the crowd. Here was a machine that didn't have a starter motor, a key or starter switch, a battery, a coil, a distributor, a rotor, or even a water pump or fuel pump, and it started more responsively and quietly than any of their engines, even their cars and pickup trucks. I had been at many an auction sale and knew that starting the engines of machines was the acid test as to the reliability of the product for sale. It was now the last item to sell, and it became almost a final tug of war. The John Deere dealer in Dawson Creek made the final and accepted bid of $1,800. It was a tractor no one would think of using, except me, and even then it brought $1,800. The people in the Peace River Country were not poor in dollars.

As an afterthought it all was a very providential event. It would be imbedded in my mind, way back though and not on the surface. It would take me over 30 more years to comprehend it and work through it, and even now it is so hard to do. I did not have my horses anymore, my cow and calf, my wagons, my thrashing machine, my John Deere D or A, and not my John Deere B. There is a song that says, "*When we wake up, life is all just a dream.*"

CHAPTER TWENTY-NINE
Leaving the Peace

The auction sale had cleared out much of our belongings, so we spent the remainder of our days in Canada rather lightly. We had kept some beds, and we did have a small cooking stove, so during the next two weeks it was just getting things ready for the move. There would be a few items that would not go this trip. We planned to return the following summer to pick up a long 16-meter wooden ladder we brought with us from Ohio. Joy's father had given it to us when we were working on the barn in Bradford, and it was useful. But we had no room for it. We would be driving the one-ton truck with its elevated stock rack, packed full. The two-wheel trailer we took with us. Ian Marchuk traded us a rack for our pickup camper shell. We felt that the camper shell was not as useful as a rack would be, so we put that on the bed of the pickup truck. We would travel with four people in the pickup cab and four in the one-ton truck cab; crowded but what we had done many times. There was a red barn that Joy's father had built for Kurt and the children when we lived in Ohio. That had gone with us in the move to Canada, and we took that back; it was sentimental to Joy and Kurt, at least. I had built a house and a barn that were modeled after our house and barn in Ohio. Those went along too in the boxcar, but we had no room for those, and they would go back with us the following summer. There were also some other smaller items that we left behind and trusted that we could get them on the next trip. That would not happen.

About two days before we left, Kurt was down by the granary, having driven the pickup down, doing something thoughtful and useful when a pickup truck skidded and went off the road right out in front of the granary. There were three men in the truck, and they asked if Kurt could pull them to get their truck back on the road again. So Kurt was accommodating to them but found out quickly that they were all three either inebriated or on drugs. As Kurt was to tell us after the event, he got one of the log chains that we did not sell and told the men to hitch it to their truck. He would turn the truck around and back up to the front of their truck and pull them back on the grade. When Kurt had backed the truck up to a position where he could pull them, he found out that all three men, roughnecks from the oil business we surmised, were incapable of attaching the log chain to their truck. They were mentally and probably physically incapable of performing that simple task. We all of us knew that strange things and people were moving around the neighborhood, and maybe our leaving was good. Kurt did, quickly and efficiently, attach the log chain to their truck and our pickup, and he pulled them out of the ditch. They left…somehow.

As I have written, this was not a celebration or reaching for a sunny, welcoming horizon. It was like going into the pit which we had experienced all our lives: a country where 1 percent of the richest rich had accumulated over 50 percent of the country's wealth and the top 2 percent had immediate control of more than 80 percent of the country's wealth. It was a surreal situation; the top 2 percent propagandizing the rest of the citizens and being successful at it. The bottom tier of citizens, financially, had mostly bought into the garbage that the rich deserved their riches. Even many middle class, hard working people, considered the US the "land of opportunity." They assumed that the rich had worked hard for it and were actually being abused and overtaxed. They had no apologies for getting 10 percent, 20 percent, 30 percent or more in reported profits, and they wanted more. I remembered back in the early and middle 1960s working in the automotive industry that at that time the Chamber of Commerce lines were that the US was where you could get wealthy and be happy and getting a profit was not a dirty word. Most anyone will agree that a company and an individual need some reward for doing work. If one spends a sum of money for doing, serving or providing a product or comfort to a customer, a bit more than the expenses is reasonably, say, 5 percent. How about 10 percent, you say? OK, but above that shows your hand.

The love of money is the root of all evil.

And also,

Money is the root of the root of all evil

It is surely easier for a camel to go through the eye of a needle than a rich man going to heaven, or earth. Reported profits is a joke, doctored up by the most dishonest to look meager. And, prophetically, it would be four months after arriving back in the US when a person would assume the office of President of the US who spoke of the "trickle down theory" of economics. What is new? This is precisely what the world has known for ages. When I heard that statement I thought of a better description: "Piss on them." It was a theory that gave the rich all they wanted, and they would spend some of it so the middle and poor people got a pittance. And it all does fit together. It does no good and one should not resist evil, but one should also not enable evil. The evildoers and thinkers are so miserable that they really do need our thoughts and ideas.

But beyond all this, mankind is evolving somehow. Gottfried Hermann visualized and wrote that mankind has four major historical phases, or epochs. It may also apply for an individual. The first epoch is the Age of Poetry, the cradle and the warm home and parents to watch us as we see beautiful sights and mystical, sensual, noble and imaginative objects. But we continue to grow, and the second age is that of the Age of Theology or the Holy. We sometimes are joined to a church, encouraged to read a religious book and attend religious services. But it cannot last, unless we completely subordinates ourselves to our fellow cultists, not a supreme being. So the third epoch is the Age of Reason or Philosophy or Enlightenment. We are reaching an age when some of our parents and friends are dying, and it requires intellectual work and a lifetime of continual learning, not something most people subscribe to. So the final age is that of Prose, or the Prosaic Age, regressing back to stories and mythologies and the cesspool of man's worst qualities. I am not sure I have arrived there, and please, reader, do not believe me, but prove what is true. But I think civilization, mankind, has arrived at the Prosaic Age where there are no longer teachers sowing calm and reason but only men sowing grains and weeds in random confusion. There is no center of gravity to prevent extreme perturbations, nor a center to attract people to gain their bearings and become human beings. Many people step forth and say they are the true leader, preaching true foolishness but couched as true wisdom. Every belief turns into blasphemy, and mysteries turn into sacrilege. The elements, the parts of the community, once evolving together and supporting one another now are locked in perpetual strife. It is chaos of the kind that was not even seen by Adam and Eve on their eviction from the Garden of Eden, no promise of fertility but of death and which God cannot create from it a world worthy of Himself. I was still hoping and enduring for me and my family.

Tätig zu Sein ist des Menschen erts Bestimmung.

Being active is the primary task of mankind.

The day of departure was sunny again. We drove the vehicles to the mouth of the lane, along Twelve Mile Road and went back to check on things, making sure we hadn't left anything important. The one-ton truck would lead the way and was parked in front with the two-wheel trailer hitched to it. Behind was the pickup truck, and both seemed to be faithfully ready to travel. We had serviced both of them and carried along a toolbox and some spare oil, just in case. And we were off....

I have throughout this reminiscence used the term *person* or *people* with careful intentions. And I know that as I write this we have over 7 billion *people* in the world. Unfortunately they are also now inhabiting the earth too. But *the meek shall inherit the earth.*

And there are, I hope, *meek human beings*, one and the same who survive after the Prosaic Age. Persons and people on the other hand are outwardly humans, but they are not, they are the Prosaic Age persona. *Humans* are sentient beings that care for other parts of creation, and I find no persons or people who care for all this. We are crude, judgmental, self-deceiving and self-indulging beings. There are signposts and guidelines for behaving as a human being, and we all have encountered them. But we don't want to recognize them or follow their directions. I like to think that some of the indigenous people, the Native Americans, the African and Afro-Americans, and the aborigines have elements of human beings. There may even be an outside chance that some whiteys qualify, but Soren Kierkegaard gave up believing that any person could achieve that condition, and Friedrich Nietzsche thought one needed go beyond people, become an Ubermensch, to reach that place. We tried by going to Canada, a rather practical, humble effort, to make this quantum leap. We tried, endured, and are still trying....

HEAVEN

Nearly every person I have talked with has expressed a desire to be in *Heaven* or some infinite place after they die. Most often this attitude is unspoken directly, but it comes through. The people who want others to know that they are "believers," people who are trying to follow one book and thereby satisfy their maker, are striving to be in *Heaven*. They even profess that they are "saved," whatever that means. But it is clear upon talking with them that they mean they are going to *Heaven*. What presumption! All religions finally try to send out one message to mankind: Be kind to one another and help them to get to *Heaven*. These religions do not say how you as an individual can get there, but they emphasize forbearance, forgiveness and kindness toward all other persons and to creation. It is better to give than to receive, and I think this means one should not even expect to get to *Heaven*. So how does the other person receiving our kindness react? Aren't they being cheated by having the "worse" part of forgiveness? It is not our call or even consideration to think about that. It does not help us or them. And of course, what is most often avoided is the corollary that *everyone* includes those we may call our enemy. It is that group of persons who we really need to show kindness toward, and the more they may display animosity, hatred or intolerance of us, the more we should forgive them. It is clear that we "should not resist evil but love the enemy." If they are hungry, feed them; if they are cold, give them your warm coat; and if they hit you, turn the other cheek. They just may get to *Heaven*. In the great tragic account of Faust, in the end Faust goes to Heaven or wherever the angels take him. With much to deny him (Faust) admission to a Heaven, as with all of us groveling in the world, somehow his striving to do the "right" things has given him redemption.

Those who are agnostic or try to be "rational" about what we are doing here finally do admit that they believe there is some sort of infinite region where we just may go. Even the atheists surely accept that they will be part of the infinite; ashes to ashes and dust to dust. It is not a stretch of truth to say that all humans either believe in a *Heaven* or want to believe it. I think most readers will admit that the picture of *Heaven* is a place where everything is **perfect** and all the problems and hang-ups that we have on earth will be gone. So, how does one know if they are in *Heaven*? There are three signs, indicators or beings who will be in *Heaven*...

Dogs

Dogs are man's best friends. They are also women's best friends. And they are children's best friends too. Sometime in the distant past dog's ancestors decided to cast their lot with humans although it doesn't really matter when or how they did it. It turned out to be a bittersweet union as many humans are not able to communicate with dogs and therefore seem to treat them badly. But dogs accept humans for themselves and forgive them more than 70 times seven. They will do anything for their master and do it to try to get their masters to *Heaven.* If a dog has a master who is mean and wants their dog to be mean, sure enough, the dog will oblige his master. So then some people say, "See, dogs are mean." But it is not the dog that is mean; it is the master. And then some people get tired of their dog, as sometimes they do even with other people, and they discard them. They often just take them far away from home and dump them. So then the dog has no one to talk with, to communicate with, or to otherwise commune with. So they find another dog or two or three and determine that they can survive without their masters. They revert to the practices of wild dogs or wolves, and humans don't like that either. So they try to destroy the "villains," not even thinking that maybe the real villain was the master who tired of his dog. And in the balance of nature, these dogs are only praying continually to their maker by doing what they were created to do. So dogs will all be in *Heaven.*

The reader might think me rash or an extremist for making such statements. But get to know a dog sometime, as a being. There are those who say dogs won't be in *Heaven,* but I say show me a person who has a soul and show me their soul. It is, in fact, superfluous to even talk about humans being the only ones having a soul. There are even statements such as *a righteous man knows the souls of his animals* and *do not deliver the soul of your turtle dove to the wicked*. So there is no reason to believe that man and woman are the only ones having a soul, or life. There is no way to know what that is, and it certainly cannot be seen or measured. But dogs are sentient beings, having feelings of joy, sadness and sorrow, hunger, pain and all the experiences you and I have. And they show those feelings by laughing, crying or depression, gratefulness for food, and gratefulness for having pain relieved. They enjoy their lives to the fullest: playing, resting, relaxing, sleeping, thinking, feeling pain and happiness, and eating. That's what we should do, to worship our makers and also to help others to see and appreciate the glories of creation. That's why dogs will be in *Heaven.*

And there are other animals that will be in *Heaven.* Cats are good beings and so are pigs, horses, cows, chickens, ducks, geese, sheep and lambs, and all the other domesticated animals. Get to know a pig sometime. They enjoy their life even if it is in a pigsty. Pigs are very clean animals, but when confined to cages they need to depend on their masters' being clean. And of course the filthiest animals on the earth are human. They foul not only their nests but everyone else's. They pollute the air, water and earth. They make noises that pollute the peace of creation and in general are not clean animals. So the pig pays for that with a dirty pen. But the pig knows Pythagorean's theorem (well known long before Pythagoras), and they use it if they can just get free of their "caretakers." They know that the shortest distance between two points is a straight line; the hypotenuse. And they take it. Dogs of course know that concept as well and so do all the animals. Only people seem to struggle with it. And the pig will be obedient if it will please his master, and he will do what he can to help the human get to *Heaven.*

So all the domesticated animals will be in *Heaven*, else why were they on the Ark?

And all the creatures that are not domesticated surely must be in *Heaven* as well. Wolves early on knew that humans were a fickle, deceiving lot, and so they did not cast their fates with the human band. But they display all the good and divine traits of their relatives; the dogs. So do bears, and deer, and elk, and moose. Tigers, and lions, and rhinoceros, and snakes, and hyenas, and giraffes. Elephants have been somewhat studied by humans, and some people do know that elephants are very family-oriented and loyal beings. It is on display if someone wants to see it.

And if you have ever fed birds in your backyard, you know that they too are going to be in *Heaven.* They are smart, energetic, loyal and sentient. The birds will all be in *Heaven* along with the fishes and other sea life. Penguins, that seem to have been pushed to those parts of the earth not wanted by anyone else, are in *Heaven.*

So if all the life will be in *Heaven*, that means also that trees, bushes, flowers and all the other plant life will be there in *Heaven*. Have you ever talked with a tree? Go into the forest sometime and listen to them. You may try to talk with them, but first listen. They are beings having feelings, and I sense when they are happy or sad. Loud noises in the forest are not pleasing to them, and I am sorry that in my life I have contributed to so much of the noise in forests. I ask the trees and plant life to forgive me.

And all of creation came from dust, the earth. So clearly rocks and sand and mud and loam and peat and coal and oil and all the creation will be in *Heaven*. All that is good (or is it GOD and GODDESS) will be in *Heaven*. Creation, if you will, is good, and so all the components of that creation will be in *Heaven.* It is not clear whether humans are part of the creation that is good. So we really need to know what *Heaven* is. Dogs are a sign that you are in *Heaven*....

Beer

Well, Beer is a beverage used to quench one's thirst. It is crisp, clean and pure when it is prepared or brewed correctly. Beer brewed as a means only to make a profit are not good, and most of them taste that way too, not good. Recipes for beer predate man's verbal or written memory. Even bread is not as old as beer. And beer does have much nutritional value besides its thirst-quenching property. Barley, hops and water. What better ingredients for food? It is abused though by humans. We seem to think of it as nothing more than an intoxicant, something to make us become drunk, zipped or buzzed. It is, therefore, abused by those who consume too much of it, and it is also abused by those who categorically reject it as unhealthy, evil or immoral. Those who believe beer to be sinful must not know that GODS gave beer to mankind as a sign of love for us. Beer is just a part of creation that can provide humans with enjoyment and encouragement. It does have the potential to be abused, just like everything else that GODS created. So we are confused and don't know how to handle it. Wine, which is another gift from the Gods, is a naturally occurring substance. Grapes, berries, fruit and grains can be prepared so that with age they make a drink that can last long past the lifetime of a human. But after a year or more just sitting as a juice it becomes this good, healthy, enjoyable drink. It has a pleasing aroma, a bountiful color, and a welcome demeanor. But, again, mankind has abused it, drinking far too much too quickly and then reacting by abstinence from it. They (mankind) even sometimes want to call it the Devil's drink and try to prohibit others from accepting this GODS' gift. Liquor, distilled wine and other refinements of wine where the alcohol content is much higher than wine are also good. Wisdom needs to be engaged (and mankind is terribly short or devoid of wisdom) in using these substances. But not only is it a concentrated and effective means for enjoyment, when used in a temperate (temperance!!), moderate way, but it also can be used externally or internally as a cleaning agent and disinfectant when some other things don't help.

And other food is going to be in *Heaven* too. All the fruits, vegetables, grasses, spices, teas, coffee, marijuana, tobacco and heroin will be there. Did I say drugs? I am sorry about that, but like all of creation, drugs, do have a place, often only for minuscule uses, but not for recreation.

Everything that has been placed here, in us or our surroundings is

Good. It is only mankind, people and person, who have abused and corrupted them. And to repeat, abuse can take two forms: either overuse or prohibition. And which of these two is the worst is hard to tell. But mankind has not been able to fit it all together correctly.

Heaven must be the place where everything fits together rightly. Any perversions, abuses or confusions are resolved there, if it is to be eternity. And as I said, beer and alcoholic beverages and food and drugs certainly have been misunderstood, perverted and abused. The extremists who want these all destroyed and even make it a crime to consume them, do have a point. But they finally miss the point further than do the abusers. So *Heaven* must have beer....

Sex

So life is defined as an ongoing process, maybe endless. And all lifeforms have a crucial process whereby they pass on their heredity to the next generations. Whether such is necessary in *Heaven* is not for us to say. But the process of passing on that heredity is sex, and it is not only the mechanics of that process but an unspoken statement by two beings that they have accepted one another for a short time, enjoy each other in many ways, and provide another way to express kindness toward each other. It surely is one of the most enjoyable ways to express feelings toward another person. But here come the hangups, as profoundly and confused as our relations to dogs and to beer. One book essentially says that all of creation is good but when man and woman pondered their relationship with each other, they were excommunicated from the good creation. I do not know what they pondered, but surely they knew that the human body is a beautiful creation, and perfect. Just like dogs and beer, though, the feelings and sensations of seeing, feeling and reacting with another human can be perverted. Again, there are standard measures for what is perverted and what is acceptable. And the nudists probably are closer to the truth on earth than those who think the human body needs to be covered. So while it is not for me to know whether there is a need to procreate and thereby practice sex for its "intended" use, I know that *Heaven* surely wants us to express our charity, compassion and love for others by all the means we have to do that with, including sex. So the monogamous relationships that are held in so high regard in many earthly circles will not suffice in *Heaven*. The women and men who are portrayed as evil because they are not afraid to share their desires and sex with many people will be in Heaven. Even the thought of *pornography* seems to be most practiced by those who have too much desire for sex; so they want to ban sex from others' lives and sometimes even in theirs. The prudes, the people who think sex is dirty and sinful, are only showing their perverted excessive dwelling on sex. These same sorts of persons are the steady reliable customers of prostitutes.

The Song of Songs in the Bible of Judaism, Christianity and Mohammed describes the celebration of loving someone, enjoying their bodies, kissing and fondling one another, and completing the relationship. And, without going off on the task at hand, the woman in that good writing was dark-skinned and probably of a different group of people from the writer. And this means that some people want to separate the different skin-colored people into separate and distinct races. Whatever a race is, the term and concept of separating people into different groups has served only to allow for intolerance of others and intolerance for creation. For an intolerance of a dog, for beer or other substances, or for creation itself is only a short hop to intolerance of other people who happen to appear different than they themselves are.

But sex is part of *Heaven*. And the process of male circumcision surely is a rite that results in man extending his pleasures with sex. The religious symbolism of this rite is just that, a symbolism. For a circumcision does not make a person closer to his creator, and it doesn't assure cleanliness either. But it does provide a way for a male member to be somewhat anaesthetized and thereby prevent its owner from climaxing in a shorter time. Whether the GODS ordained such practice of circumcision is open. I know that an un-circumcised male surely can enjoy sex as well, and maybe better, than the circumcised male and thereby celebrate the God's creation of a beautiful woman or man.

Sex is one of the most intimate, effective and truthful ways of expressing love, care and respect for another being, when done in wisdom. And in the maze of thick entangled writings from the past and the present of those authors who can't handle the subject or concepts of sex, one is left with the thought that to abstain or ignore sex is good. Heaven forbid this! Such fools, maybe demonic beings, who want others to think this are surely not going to be in *Heaven.* Through all of one's life, the thought of sex and the possibilities for it to occur, and it actually occurring, are there. It seems that only in *Heaven, or earth,* will it all be correct and the distortions and dilemmas of man toward creation will be eliminated.

For the meek shall inherit the earth.

Kurt Rolle was born in northeastern Wisconsin along Lake Michigan in 1938. From the age of 3 he lived with his family in east central Indiana, graduating from Winchester High School in 1956, Purdue University in 1960, and earning a Ph.D. in Mechanical Engineering from the University of Dayton in 1978. He spent many of his early summers in northern Wisconsin with his grandparents in a rural setting.

In 1961 he married Joy Fruit and they have had six children. Since 1980 Kurt has lived on a small farm in the Driftless Region of Southwestern Wisconsin, teaching mechanical engineering at the University of Wisconsin at Platteville, doing engineering consulting, small scale farming, and research on wind, solar, geothermal, and all forms of sustainable energy and life styles.

You can contact Kurt at sandbergprarie@yousq.net.

CPSIA information can be obtained at www.ICGtesting.com
Printed in the USA
LVOW042040120612

285714LV00002BA/3/P

9 780984 653614